STUDENT STUD

MW00906167

to accompany

STATISTICS

PRINCIPLES AND METHODS

Second Edition

RICHARD A. JOHNSON
GOURI K. BHATTACHARYYA
University of Wisconsin at Madison

JAMES C. CURL
Modesto Junior College

JOHN WILEY & SONS, INC.
NEW YORK • CHICHESTER • BRISBANE • TORONTO • SINGAPORE

Copyright © 1992 by John Wiley & Sons, Inc.

All rights reserved.

Reproduction or translation of any part of
this work beyond that permitted by Sections
107 and 108 of the 1976 United States Copyright
Act without the permission of the copyright
owner is unlawful. Requests for permission
or further information should be addressed to
the Permissions Department, John Wiley & Sons.

ISBN 0-471-55549-5

Printed in the United States of America

10 9 8 7 6 5 4 3 2

Printed and bound by Malloy Lithographing, Inc.

Preface

...for there is nothing either good or bad,
but thinking makes it so.

– HAMLET Act 2, Scene 2

The preface would usually tell you what to expect in the study guide. However an entire chapter, Philosophy of The Study Guide, has been written to explain what is in the study guide and how to use it with Johnson and Bhattacharyya's <u>Statistics: Principles And Methods, 2Ed</u> . I encourage you to take time to read through the first chapter of the study guide to learn about the things that you can do to help you do well in statistics.

This study guide reflects the commitment, hard work, and help of several people. I would like to use the preface to express my thanks to three people who helped me translate the ideas I had about the study guide into reality:

- Brad Wiley II, editor at John Wiley and Sons, for his vision, encouragement, and guidance.
- Joshua Delahunty, illustrator and typographer, for his untiring commitment and extraordinary talent in desktop publishing in helping produce the study guide.
- Heather Curl, for her ongoing critical reviews and copy editing of the many revisions of the manuscript.

They shared in what has been both a challenging and rewarding experience, creating a unique study guide that is easy to read and use. To the extent that you find the study guide helpful, they each deserve much of the credit.

Finally, I would like to thank the students in my statistics classes, both past and present. Their questions, discussions, and involvement in learning statistics provided the inspiration for this study guide.

Contents

chapter zero

PHILOSOPHY OF THE STUDY GUIDE

A good notation
has a subtlety and suggestiveness
that at times
makes it like a live teacher.

– Bertrand Russell

PHILOSOPHY BEHIND THE STUDY GUIDE

The philosophy of this study guide is to help students understand and use inferential statistics. At this point in the study guide, it may be ambitious to expect the reader to feel comfortable with the stated philosophy. It may help to first explain the structure of the text as it relates to inferential statistics. The text is subdivided into three main areas: descriptive statistics, probability, and inferential statistics.

Descriptive statistics is presented in the first section. The calculated values from this section are called sample statistics. The arithmetic mean is a familiar example. The sample statistics are used to challenge accepted or hypothesized values of the population from which the sample was obtained. Presenting and evaluating the challenge is called hypothesis testing and is a major part of the third area.

Descriptive
Statistics

The second area of the text involves **probability**. A familiar example from probability involves the chance of getting a head from one toss of a coin. The section on probability introduces concepts that are very important for a better understanding of inferential statistics.

In particular, probability is used:
- To explain how the data we collect is distributed.
- To determine if our sample statistic is a rare event.
- To find the size of the error that can result if we reject the accepted value from the population.

The third area presents **inferential statistics**. Hypothesis testing, a central theme in statistics, is what we typically refer to as inferential statistics. The important point to remember is that inferential statistics incorporates everything we study in the course. It is where all the concepts come together. More importantly, the concepts in inferential statistics are essential to using statistics in applications relating to business, science, medicine, industry, education, research, and many other fields.

The problems students have with statistics often occur at what we will call "Hot Spots". These Hot Spots are concepts that may be difficult to understand or topics that simply require more help than is typically provided in the text. Sometimes it is just a matter of needing more practice so that the concept, formula, or notation is understood. The Hot Spots are designed to provide the additional help and are the focus of this study guide.

One final point needs to be made regarding the philosophy of this study guide. Every effort has been made to make the study guide readable. There is a generous use of space to allow the reader to take notes, respond, and to interact with what has been written. Please do highlight it, mark it up, write in it, and get involved.

THIS STUDY GUIDE ASSUMES

- The focus should be on difficult areas called Hot Spots.
- It is important to understand what you are doing with formulas and why you are using them.
- The use of notation will increase your understanding.
- There is a need to discuss the concepts in a study group to increase understanding.
- You are willing to ask for help from the instructor as needed.

THE CONTENTS OF EACH CHAPTER OF THE STUDY GUIDE

- **INTRODUCTION**
- **HOT SPOTS**
- **REVIEW**
- **DISCUSSION QUESTIONS**
- **TEST**
- **TEST QUESTIONS AND ANSWERS**

OVERVIEW OF THE STUDY GUIDE

Each chapter of the study guide begins with a quotation and an **INTRODUCTION**. The introduction gives a concise overview of the concepts and key terms in the chapter. The content of the chapter is related to the broader picture of inferential statistics that encompasses the subject from the first to last chapter of the text. The important terms of the chapter are highlighted. An important part of the philosophy of the study guide is that the user will become very involved. It is hoped that you will take time on your own, and if possible spend time in your study group, discussing the definitions, formulas, and notation that is part of each chapter..

Introduction

The next section of the study guide is a list of **HOT SPOTS**. The Hot Spots, the major feature of the study guide, focus on the statistically complex areas of each chapter. Each Hot Spot, designated with the symbol ❋, is keyed to a page in the study guide, and the authors' last names, Johnson and Bhattacharyya, are used to indicate the related sections of <u>Statistics: Principles And Methods, 2Ed</u> . You are encouraged to find other Hot Spots and use them as a focus for the discussions in your study group.

Review

The **REVIEW** section consists of the Hot Spots. It is the most important part of the study guide. Each Hot Spot discusses a problem area of statistics and includes examples, and when appropriate, sample problems with complete solutions, and sample problem-and-answer selections that provide additional drill for each problem area.

Discussion Questions

A set of ten **DISCUSSION QUESTIONS** follows the review section. The discussion questions focus on the statistical concepts in each chapter. They can be used as a basis for discussion in your study group or for individual reflection. You are encouraged to meet with your instructor if you have trouble answering the discussion questions.

Test

A ten question **TEST** follows the discussion questions in chapters 2 through 15. The test questions provide additional problems that can be used in preparation for exams. These problems are written with minimal notation, stressing verbal statements that encourage the reader to make decisions about which formula and notation to use.

Test Questions And Answers

The **TEST QUESTIONS AND ANSWERS** section is at the end of the chapter. The test questions are repeated for your convenience, and complete solutions are given for each of the test problems.

HOW TO USE THE STUDY GUIDE

Naturally, each student may use the study guide in a different way. Whatever you do, interact with the study guide. Write in it, add your own personal notes, and make it work for you. The following suggestions may help you make better use of the study guide.

Before the lecture, it is suggested that you read the Introduction section and focus on the key terms that are highlighted. The space in the margins can be used to add notes from your reading of the text and related information from the lecture.

The section on the review of Hot Spots should be used when you are doing homework or are involved in a discussion group. The Hot Spots are listed by title and cross referenced to chapter sections in Statistics: Principles And Methods 2Ed. You can also reference a Hot Spot by looking for the Hot Spot symbol, ✳, in the margin of the study guide. The Hot Spots are intended to provide additional explanation of difficult areas. The sample problem-and-solution and problem-and-answer selections that are found in the Hot Spot reviews can be used to both reinforce the ideas presented in the Hot Spot and provide more practice for problems on tests.

Use this margin space to work out your answers to problems in the text.

Although the discussion section is useful in group discussion, it is even more important for individual reflection. Taking time to mentally answer each question provides a good opportunity to see how well the material is understood. If at all possible, try to verbalize the answers to the questions to get an even better feeling for the material.

The chapter test and the related answer section can be used for both test preparation and a general review of the material in the chapter. If more than one chapter is included on an exam, the problems from the chapter tests involved can be divided up to create two or three typical exams. These trial exams can be done in a restricted time frame to provide practice under pressure to prepare for the class exam.

STRATEGIES FOR SUCCESS

Students often want to know if there is some way they can guarantee that they will pass statistics. They want to know what course or courses they should first take, how they should study, what it will take to pass the course, etc. These questions will be answered as you read through this section of the study guide.

The Nature of the Problem

Many authors and instructors alike believe that statistics is not a respecter of background. They argue that both beginning lower division students and graduate students are often equally frustrated when taking statistics. To some extent their statements are valid.

Students with minimal backgrounds in quantitative areas will find the formulas, notation and number crunching uncomfortable. The use of the computer to minimize the trauma of number crunching often creates yet another trauma of learning how to use a computer. On the other hand, the student with a strong math or quantitative background is often frustrated with the language of statistics and the somewhat uncomfortable world of discussion and interpretation.

The reason for the problems most students experience is the dual nature of statistics. The course involves two different areas of thinking. The first involves the verbal area and depends heavily on your ability to learn concepts, discuss ideas, and interpret results. The second involves the quantitative area that depends heavily on notation, formulas, and calculations.

Many students find that they have strong abilities in only one of these two areas. It is rare to find students that are equally strong in both the verbal and quantitative areas. Thus, the math and science students are bothered by the excessive use of language to discuss concepts. The non-math and non-science students get bothered with the formulas and number crunching.

There is a lot that can be done to help what may appear to be a hopeless situation. The following suggestions will help.

Study Groups

The first thing to realize that is you are not alone in your feeling of frustration. There are many other students feeling equally unsure of themselves. But this is the first place you can make a difference. You will find that if you form a small study group so that you can talk not only about your frustration but also about the problems you are having in the course, you will be less frustrated. You will find that you will better understand the material.

The key point is that you must participate in the group. You need to verbalize what you are having trouble understanding and try to explain to others what you understand. You will be amazed at the difference the study group makes. You no longer are sitting alone late at night fighting statistics and getting more and more upset. Instead, you have help if you are stumped. You will also find that you can learn in more depth as you help someone else who is having difficulty understanding. Studying alone usually results in you getting angry with the subject, the instructor, the book, and yourself...not in what should be happening; you learning statistics.

Of course, there is more to success than just forming a study group; but it is an important beginning. As a group you may want to invite your instructor in occasionally to help explain an idea that everyone is having problems with. Your instructor may be able to provide you with some additional sources for problems to do or some practice tests that you can use to prepare for exams. You may want to go together and pay for a tutor to work with you as a group for few hours each week. But most important, you just need the mutual support the group provides as you do homework and prepare for tests.

Before we leave the topic of support groups, it is important to add a comment about how valuable a phone call can be when you are stuck on a problem. A short call to someone in the class can often help you see what you need to do. All you need to do is get the phone number of one or two people in your class and make an agreement as to when you can call each other. It may help to do your homework early enough in the evening so that if you have to make a call, you can do so without waking up the other person.

One of the problems with creating either a study group or a phone network is meeting other students who are in the class. One simple way of dealing with the problem is to prepare a 3 × 5 card that indicates your interest in forming a study group and lists the times and days you would like to meet. The same could be done with a phone network, and interested students could then meet at the end of a class session to compare interests and time. If nothing else, you could ask the instructor to announce that there is an interest in forming a study group, and for those wanting to get information about the study group to meet briefly after the class.

Computer Readiness

Another thing that you can do to help yourself be successful in statistics is to learn how to operate a computer. You can then use the statistical software package, Minitab, that is referenced in many of the of the chapters of <u>Statistics–Principles and Methods 2Ed</u>. Most colleges and universities have introductory courses in the use of a computer. You do not need to learn much more than how to use the computer keyboard, a few commands, and some of the language that seems to be such a part of working with computers. You will be amazed how much just a simple background in typing can help ease the frustration of using a computer. The computer can be a real asset when it comes to actually working with numbers. Your instructor may involve you in a computer lab to do some of your work. Time spent getting familiar with the computer will pay off.

A Study Guide For Statistics

If your statistics course does not involve a computer lab, you may find that it is worthwhile to spend some time with the new student edition of the statistical software package EXECUSTAT® (from the developers of STATGRAPHICS®). The new student edition of EXECUSTAT provides excellent on-screen help and an interpretive section that will make using the computer a pleasurable learning experience.

Use Of Calculators

Although you may decide not to use a computer, you really need to invest in a calculator to help minimize the time that you spend doing calculations. There are many calculators available in a variety of price ranges. However, calculators that have a statistical mode are by far the most valuable to you for use in this course.

If possible, choose a calculator that (1) allows you to find reported P values for hypothesis tests using the z, t, F, and χ^2 distributions without fighting the tables in the text, (2) contains programs that allow you to find both individual and cumulative probability values for the binomial distribution, and (3) gives the correlation coefficient and the equation of the regression line for bivariate data.

At the time of this writing, the most useful calculator for statistics in the under-$50 range is Hewlett Packard's HP-21S. It is easy to use and complements the text throughout the course. The HP-21S was used in the study guide. Unfortunately, Hewlett Packard recently announced plans to discontinue production of the HP-21S. Whichever calculator you choose, you need to spend some time reading the calculator manual and learning how to use the calculator before starting the course, or at least during the first few weeks. You will then find that it will be much easier to work with the calculator during the semester or quarter.

EXECUSTAT is a registered trademark of Strategy Plus, Inc.
STATGRAPHICS is a registered trademark of Statistical Graphics Corporation and an STSC Plus*Ware software product.

WHAT COURSES DO I NEED BEFORE STATISTICS?

Most statistics texts assume you have a good background in algebra. It is difficult to survive statistics with less than a background in intermediate algebra. Two years of high school algebra would be the equivalent of what most colleges identify as intermediate algebra. It goes without saying that you need to do well in such courses. A grade of B or higher provides a reasonable guarantee of having a strong enough background. A grade of C, particularly if on the low end of the scale, might not be a strong enough background. However age, maturity, commitment, and good study habits are often as important as the prerequisite course and grades. If you feel you are over your head, take time to visit with your instructor during an office hour. Talk with the instructor about your concerns. The time you spend with the instructor usually provides an answer to the question of whether or not you are prepared to take statistics.

WHAT DO I WANT FROM THE COURSE?

Very often students take statistics because it is a required course for their majors. They simply see the statistics course as another obstacle to overcome on the road to a baccalaureate degree. What most students do not realize is how important the statistics course will become to them if they decide to pursue graduate work. And for those who do not go on to graduate school, many are surprised to learn how they end up using statistics in their job. The point is that you have two choices you can make. You can take statistics as what is often called a "cook-book course" or you can take the course with the goal of understanding statistics.

Students taking the course using the cook-book approach often find themselves overwhelmed with the notation and the number of formulas that must be used and, at times, memorized. Such students do not have any way of organizing the information other than a brute force memory related process. If instead, you approach the course from the perspective of understanding inferential statistics, the formulas can be viewed in a much more constructive manner. For example, the formulas used for what is called a test statistic simply become variations of one simple formula. What was disorganized and overwhelming becomes easier to use and remember.

When the course is viewed from the perspective of understanding inferential statistics, the entire course can make sense and each chapter is a logical extension of the earlier material. You as the student can use this extension as a basis for asking how each new concept adds to the overview of what is being done in statistics. This view of *understanding* statistics, rather than using the cook-book approach is very consistent with the earlier suggestion about getting in a study group to help you not only succeed but *understand* the course itself.

HOW MUCH TIME DO I NEED TO SPEND ON STATISTICS?

Clearly the amount of time each student needs to spend on a statistics course is a very individual decision. The demands of the instructor will be a factor as well as the student's background and general abilities in academia. However, there are some general considerations that are worth discussion.

You must spend enough time to do more than just a "once over" on the problems that have been assigned by the instructor. You need to have time to read the material in the text. There then must be time provided to talk with others in a study group, or anyone else that will listen — even if it is just the wall of your study area. You need to spend time organizing the new formulas and verbalizing the new vocabulary in addition to doing the problem sets assigned. Beyond this, there is the need to study for exams. Assuming that most statistics courses meet three hours a week, the old rule of two hours of study for each hour of lecture seems like a good place to start. Additional hours should be added on for test preparation and study group discussions. A time frame of between 6 and 10 hours per week is realistic in most situations. Obviously, some students may get by with less time, and many students will spend more than this amount of time. A good rule of thumb would be to spend approximately 5 hours minimum during the week on homework and reading and an additional 3 hours each weekend in a study group.

A final point relative to the question of how much time to spend on statistics is in regard to the frequency of how often one should study. There is a great difference in studying one hour each day when compared to doing the five hours in one day. For example, a marathon session on a Saturday will not be as productive as working on statistics on a daily basis.

REVIEWING YOUR MATH BACKGROUND

*Notation,
Tools,
Concepts,
And Their
Applications
To Statistics*

One of the most important things you can do to help yourself is to make sure you review the material from the algebra courses that you took as a prerequisite for the statistics course. You may also have taken a geometry course that will provide insight into hypothesis testing. Not everything in these courses needs to be reviewed, but there are some key areas that will help. You do not need to tackle all these areas at one sitting. But as you find yourself getting ready to cover the chapter involved it would pay to take time to review the related material first.

The review areas that follow are related to the specific chapter(s) in Statistics: Principles And Methods, 2Ed (as indicated with **Stat:**). Chapters referenced in algebra and geometry are also listed (as indicated with **Alg:** or **Geo:**, respectively).

1. **Subscripts** are used with variables, as in x_1, x_2, x_3 to represent different variables in data analysis and in formulas for sample statistics and population parameters.
   Alg: Reference the chapter on sequences and series.
   Stat: Used in all the chapters excluding the Introduction.

   Subscripts

2. The symbol Σ (the **Greek letter sigma**) is used to indicate a sum, as in $\sum_{i=1}^{3} x_i = x_1 + x_2 + x_3$.
   The symbol Σ is used in the definition of sample statistics and population parameters such as the mean and variance.
   Alg: Reference the chapter on sequences and series.
   Stat: Used in all the chapters excluding the Introduction.

   Greek Letter Σ

$\sum x$ *versus*

$\sum x^2$ *versus*

$\left(\sum x\right)^2$

3. The difference between expressions like $\sum x, \sum x^2$, and $\left(\sum x\right)^2$ is important in finding values for sample statistics like the mean, variance, and slope and intercept of the regression line.

    Alg: Reference the chapter on sequences and series.

    Stat: Used in Chapter 2 on Organization And Description of Data...Chapter 3 on Descriptive Study of Bivariate Data ...Chapter 12 on Regression Analysis...Chapter 15 on Analysis of Variance.

Square Versus Square Root

4. The difference between the **square** (x^2) **and square root** $(\sqrt{x})$ of a number is important to understanding the difference between the standard deviation and variance and the correlation coefficient and the coefficient of determination.

    Alg: Reference the chapter on radicals.

    Stat: Used in Chapter 2 on Organization And Description of Data...Chapter 3 on Descriptive Study of Bivariate Data...Chapter 6 on Probability Distributions...Chapter 7 on The Normal Distribution... Chapter 8 on Variation in Repeated Samples–Sampling Distributions.

Factor Tree

5. The idea of a **factor tree** to write a composite number as a product of primes is used to create tree diagrams in probability.

    Alg: Reference the chapter on least common denominators.

    Stat: Used in Chapter 4 on Probability.

6. The idea of a **proportion**; as in a fraction or comparative share of the whole; is used in grouped frequency tables, probability, and hypothesis testing. This proportion is more like a ratio and different from the way a proportion is defined in algebra as the equality of two fractions.

   Alg: Reference the chapter on ratio and proportion.

   Stat: Used in Chapter 2 on Organization And Data,...Chapter 4 on Probability...Chapter 9 on Drawing Inferences From Large Samples... Chapter 11 on Comparing Two Treatments...Chapter 14 on Analysis of Categorical Data.

   Proportion

7. The meaning of a **variable** and the idea of a **domain** for the variable is used to define probability distributions for both discrete and continuous variables.

   Alg: Reference the chapter on variables and functions.

   Stat: Used in Chapter 5 on Probability Distributions...Chapter 6 on The Binomial Distribution And Its Application In Testing Hypotheses...Chapter 7 on The Normal Distribution.

   Variable Versus Domain

8. The use of $f(x,y)$ notation to represent a **function** of more than one variable (as in $f(x,y) = 5x^2y^3$) in finding probabilities for the binomial distribution.

   Alg: Reference the chapter on functions and functional notation.

   Stat: Used in Chapter 6 on The Binomial Distribution And Its Application In Testing Hypotheses.

   Function

9. The idea of an **inequality**, like $x \leq 3$ and the related graph on the number line, is used with function notation to indicate a desired probability with both the binomial and normal distributions.

   Alg: Reference the chapter on linear inequalities.

   Inequality

Stat: Used in Chapter 5 on Discrete Random Variables And Their Probability Distributions...Chapter 6 on The Binomial Distribution And Its Application In Testing Hypotheses ...Chapter 7 on The Normal Distribution.

Pascal's Triangle

10. The expansion of $(p + q)^n$ using **Pascal's triangle** is used to find probabilities for the binomial probability distribution.

   Alg: Reference the chapter on the binomial expansion.

   Stat: Although Pascal's triangle is not used in the text, it can be used in Chapter 6 on The Binomial Distribution And Its Application In Testing Hypotheses.

Use Of A Table

11. The **use of a table** to indicate values of a function is used to define discrete probability distributions. The tables are also used to help in the computation of statistics like the mean, variance, correlation coefficient, the slope and intercept of the regression line, and the graph of the regression line.

   Alg: Reference the chapter on functions and graphing.

   Stat: Used in all chapters excluding 1 on Introduction.

Area Under A Curve

12. The **area under a curve** is used to find probabilities for continuous variables.

   Alg: Reference the chapter on formulas for areas.

   Stat: Used in Chapter 7 on The Normal Distribution.

Transposing A Vertical Axis

13. The concept of **transposing a vertical axis** using the substitution $x = x - \mu$ to create a standard normal curve that is symmetric to the vertical axis.

   Alg: Reference the chapters on completing the square.

   Stat: Used in Chapter 7 on The Normal Distribution.

14. The graph of the **exponential function** $f(x) = e^{-x^2}$ is used to
    define the standard normal curve.
    Alg: Reference the chapter on exponential functions.
    Stat: Used in Chapter 7 on The Normal Distribution.

    Exponential Function

15. The idea of **symmetry of a curve** like $y = x^2$ is used to simplify
    using a table .
    Alg: Reference the chapter on graphing second degree functions
             in two variables .
    Stat: Used in Chapter 7 on The Normal Distribution.

    Symmetry Of A Curve

16. The graph of the **linear function** $y = mx + b$ using the slope
    intercept form of graphing is used in bivariate data analysis.
    Alg: Reference the chapter on graphing linear equations.
    Stat: Used in Chapter 10 on Simple Linear Regression.

    Linear Function

17. **Greek letters** are used in statistics to represent special values. The
    letters α (alpha) and β (beta) are used to represent probabilities
    in hypothesis testing and the letters μ (mu) and σ (sigma) are
    used to represent values of the population parameters.
    Geo: Reference the chapter that introduces notation.
    Stat: Used in all chapters excluding chapters 1, 2,4, and 16.

    Greek Letters

18. The idea of a **hypothesis** from geometry is used to create a null
    and alternate hypothesis in hypothesis testing.
    Geo: Reference chapter on creating hypotheses.
    Stat: Used in Chapter 6 on The Binomial Distribution And Its
             Application In Testing Hypotheses...Chapter 9 on Drawing
             Inferences From Large Samples.

    Hypothesis

19. The idea of proving something in statistics is very similar to **indirect proofs** in Geometry where the assumed hypothesis leads to a contradiction and is thus rejected in favor of the alternate position. In statistics if the assumed hypothesis leads to a rare event, then the hypothesis is rejected in favor of the alternate.

    Geo: Reference the chapter on the logic of proofs.

    Stat: Used in Chapter 6 on The Binomial Distribution And Its Application In Testing Hypotheses...Chapters 9 through 16 that involve hypothesis testing.

20. In Geometry, **if the hypothesis leads to a false result**, the hypothesis is wrong. In statistics, if the hypothesis leads to an unlikely result, the hypothesis is probably wrong.

    Geo: Reference the chapter on hypothesis testing and the logic of proofs.

    Stat: Used in Chapter 6 on The Binomial Distribution And Its Application In Testing Hypotheses...Chapter 9 on Drawing Inferences From Large Samples.

$$\Sigma_\circ$$

chapter one

INTRODUCTION

Some people hate
the very name of statistics,
but I find them
full of beauty and interest.

— Sir Francis Galton

INTRODUCTION

We start our study of **statistics** in this chapter with several
definitions. The definition of statistics is divided into two areas,
descriptive and inferential. We define **descriptive statistics** as the
branch that includes collecting, organizing, displaying, and describing
data. In contrast, **inferential statistics** is defined as the branch that uses
sample results to help make decisions or predictions about populations.

We will be working with **data** that we collect from what is called
a **population**. The population consists of all elements whose
characteristics are being studied, and the portion of the population we
select for study is called a **sample**. The sample is simply a collection of
some of the data from the population. The data itself can either be
quantitative (can assume any numerical value) or qualitative (can not
assume a numerical value.

We will use the sample data to calculate a statistic. A **statistic** is
defined as a numerical property of the sample. The test average that we

determine in each of our courses at the end of a quarter or semester is a familiar example of a statistic. The sample statistic is used as an estimate of a population **parameter**, a numerical property of the population. The accuracy of our estimate depends on how well the sample represents the population. As such we are concerned that a sample is a **representative sample** that contains the characteristics of the population.

Both the sample statistic and the population parameter become an important focus in inferential statistics. Generally speaking, we use the sample statistic to challenge an accepted value of the population parameter. If the statistic and parameter differ greatly, we argue that the assumed value of the parameter should be rejected. Much of our work in the following chapters is an effort to quantify the phrase "differ greatly". Again our decision will depend on how representative of the population the sample is.

There are several methods used to obtain samples. The method that is important to later work is called a **random sample**. In a random sample from the population each element of the population has some chance of being selected. If each element has the same chance of being selected, we call the sample a **simple random sample**.

At times, there are limitations on the resources available for obtaining a simple random sample. We then may have to choose between several other samples that involve a chance process to get the sample, but do not qualify as simple random samples. The stratified sample, cluster sample, and systematic sample can then be used. The **stratified sample** involves dividing the population into groups we call strata. We then take a sample from each strata. The **systematic sample** involves taking every nth piece of data from a list of the population; whereas, the **cluster sample** involves taking a random sample from some selected group (cluster) of the population. One of the most common samples is called a **sample of convenience**. It is a sample that already exists and is available for study, but the elements are not chosen by a chance process.

CHAPTER 1 HOT SPOTS

1. **Descriptive Statistics Versus Inferential Statistics.**
   Starts on **page 1-4.** J&B 1.3

2. **A Sample Versus A Population.**
   Starts on **page 1-5.** J&B 1.4

If you find other HOT SPOTS, write them down and use them as a focus of your discussions in the study group. Or you can use the HOT SPOT as a the topic for a help session with your professor.

*Johnson &
Bhattacharyya
Section 1.3*

Hot Spot #1 – Descriptive Statistics Versus Inferential Statistics

There is an easy way to understand the difference between descriptive statistics and inferential statistics. We simply look at the root word in each case. Describe is the root of descriptive, and infer is the root of inferential.

In descriptive statistics, we are trying to summarize and describe the prominent features of data collected from the population. We do this by either calculating sample statistics as estimates of what we will later define as population parameters or through the use of graphs that show how the data is distributed. In inferential statistics, we evaluate the information and draw conclusions about the population parameters based on the analysis of the data. An important part of inferential statistics is an assessment of the uncertainty involved in the inference.

A very easy way to distinguish between a statistic and parameter is to remember that the word sample starts with an s just like statistics starts with an s. In the same way, both population and parameter start with the letter p. In fact, it might be a good idea to get into the habit of saying the phrases "sample statistic" and "population parameter" rather than the words statistics and parameter. Admittedly, this oversimplifies the problem, but a little association may go a long way to keeping the terms in perspective. On a more serious level, we need to look at the relationship between a sample statistic and a population parameter in both descriptive and inferential statistics.

In the branch of descriptive statistics, it is important to understand that we seldom know the value of a parameter. We are trying to estimate the parameter which is a measurable characteristic of the population. Since most populations are quite large, it would be difficult, if not impractical, to measure a characteristic of the population. Therefore, we use the sample statistic as an estimate of the population parameter. In the beginning, we will call this estimate a point estimate. Later on, we will extend our concept of an estimate to what we will call an interval estimate.

We have a different situation in inferential statistics where we assume a value of the population parameter. This is part of the inferential area labeled hypothesis testing. The hypothesis is that the population parameter is equal to the assumed value. The sample statistic is calculated and compared to the hypothesized value of the population parameter. If the sample statistic differs greatly from the assumed value of the population parameter, we say that we reject the hypothesis. Much of our work in the following chapters is an effort to quantify the process of hypothesis testing. As we do so, it becomes very important to distinguish between the value of the sample statistic and the value of the population parameter.

Hot Spot #2 – A Sample Versus A Population

✳
Johnson &
Bhattacharyya
Section 1.4

We define a population as the set of measurements corresponding to the entire collection of units for which inferences are to be made. Usually, we are interested in some measurable characteristic in the form of numerical data. When it is possible and practical, we study populations directly. The US Census is a familiar effort to collect data from an entire population. But most of us do not have the resources of the federal government available to do a census. So, we must settle for collecting smaller sets of data.

This smaller collection of data is the sample. Ideally we want as large a sample as our resources allow to insure that the sample is representative of the population. We also want to avoid any bias that might result if only certain parts of the population are selected. We accomplish this by using a probability sample where the sample is obtained by a chance process. Most of the data analysis we do later on assumes that our sample is a random sample where all samples of the same size have an equal chance of being selected.

CHAPTER 1 DISCUSSION QUESTIONS

These questions may be used in your study group or simply as topics for individual reflection. Whichever you do, take time to explain verbally each topic to insure your own understanding. Since these questions are intended as topics for discussion, answers to these questions are not provided. If you find that you are not comfortable with either your answers or that your group has difficulty with the topic, take time to meet with your professor to get help.

1. What is meant by the word statistics?

2. What does scientific inquiry refer to?

3. What is the role of experimental design?

4. What is the difference between descriptive statistics and inferential statistics?

5. What is the difference between a sampling unit and a sample ?

6. What is the difference between a population and a sample?

7. What are three common characteristics of studies that are done in statistics?

8. What are the major objectives of statistics?

9. What characterizes a good design in the sampling process?

10. What is the difference between the early use of statistics and the modern role of statistics?

$$\sum \circ$$

chapter two

ORGANIZATION AND DESCRIPTION OF DATA

"Data! data! data!"
he cried impatiently.
"I can't make bricks without clay!"

– Sherlock Holmes

INTRODUCTION

In this chapter we begin working with data. The data may be
qualitative or **quantitative** and come from either what is called a
population of data values or simply a subset of the population that we
call a **sample**. We will generally use the letters x or y to represent the
unspecified data values. The idea of frequency will be used to both
summarize and organize a collection of data values. We accomplish this
by creating frequency tables and drawing graphs.

We obtain a **frequency table** by listing the possible data values
together with the frequency, or number of times the data value is found
in the collection of data. The result is called a **frequency distribution**.
We will focus on the idea of the frequency distribution throughout the
remaining chapters of the book.

There is a second type of frequency distribution we create by first
dividing the collection of data into intervals called **classes**. We then find

the number of data values in each class. A large collection of data can be reduced to a smaller number of classes to get a better idea of how the data is distributed. We call the result a **grouped frequency distribution** where **class limits** are used to create the classes.

We use graphs to summarize and organize data collections. The **histogram** involves drawing a bar graph where the bars are contiguous (touching) and represent the **frequency** or **relative frequency** of the data in each class of the grouped data frequency distribution. If we use a dots to indicate the frequency of each interval, the diagram is called a **dot diagram**.

We will also work with the **stem-and-leaf display** and the **boxplot** from the recently developed field of EDA (exploratory data analysis). We create the stem-and-leaf display as we record data, and the resulting display resembles a histogram. The **boxplot** provides a graphic display of the numerical features of a collection of data.

In this chapter we begin our descriptive analysis of data. We focus on measures of center and variation. The mean, median, and mode are three measures of center. The **mean** is the usual arithmetic average; the **median** is the middle value in a ranked data list; and the **mode** is the data occurring most often. The range, variance, and standard deviation are the three measures of variation that are used. The **range** is the highest data value minus the lowest data value; the **variance** is the modified average of the squares of the deviations of the data values from the mean; and the **standard deviation** is the square root of the variance.

The mean and standard deviation are very important to the work we will do in later chapters. We see the beginnings of this in the current chapter as we work with the **Empirical Guideline** that applies to bell shaped frequency distributions. The Empirical Guideline tells us how often we can expect to find data in an interval about the mean. This ability to quantify how often data should be found in an interval is at the heart of inferential statistics.

CHAPTER 2 HOT SPOTS

1.   **Number Of Intervals In A Grouped Frequency Table.**
     Starts on **page 2–4.** Problems on **pages 2–5, 2–6.** J&B 2.3

2.   **Frequency Versus Relative Frequency.**
     Starts on **page 2–6.** Problems on **pages 2–8, 2–9, 2–10.**
     J&B 2.3

3.   **Value Of Median Versus Location Of Median.**
     Starts on **page 2–11.** Problems on **pages 2–12, 2–13.** J&B 2.4

4    $\sum x^2$ **Versus** $\left(\sum x\right)^2$.
     Starts on **page 2–13.** Problems on **pages 2–15.** J&B 2.5

5.   **Checking To See That We Have** s **Rather Than** s^2.
     Starts on **page 2–16.** Problems on **pages 2–17, 2–18.** J&B 2.5

6.   **The Meaning Of** $\bar{x} \pm ks$.
     Starts on **page 2–18.** Problems on **page 2–19, 2–20.** J&B 2.5

7.   **When** P_k **Does Not Result In** $k\%$ **Of The Data Below** P_k.
     Starts on **page 2–20.** J&B 2.4

8.   **Rounding Off.**
     Starts on **page 2–22.** J&B 2.4, 2.5

If you find other HOT SPOTS, write them down and use them as a focus of your discussions in the study group. Or you can use the HOT SPOT as the topic for a help session with your professor.

❋

Johnson &
Bhattacharyya
Section 2.3

Hot Spot #1 – Number Of Intervals In A Grouped Frequency Table

We often have a large collection or set of data. It is then difficult to see how the data is distributed. We try to simplify the data by dividing a large collection of data into a smaller number of intervals. The argument is that it is much easier to examine the shape of a distribution when working with between 5 and 15 intervals than with a larger number of data. Of course, this brings up the issue of what is meant by a large number of data. If there are less than 30 pieces of data, there really is not much of a need to regroup data. Clearly, there is a need to regroup the data if there are more than 100 pieces of data. It can be confusing for the beginning statistics student when trying to choose the appropriate number of intervals for larger data sets.

There is no absolute Guideline that needs to be followed in selecting the number of intervals. However, there are several different guidelines that can be followed. One of the simplest Guidelines is to take the square root of the sample size to find the number of intervals to use. This Guideline works fairly well for data sets that are larger than 30 and smaller than 150. For example if the collection had 50 pieces of data in it, we would use 7 intervals.

$$\text{Number of intervals} = \sqrt{50} \approx 7$$

Unfortunately this Guideline tends to give more intervals than is really necessary when we work with very large data sets.

A more effective Guideline uses the powers of 2 to find the number of intervals. We first need to make a reference list of the powers of 2:

A Study Guide For Statistics

The First Ten Powers of 2	
$2^1 = 2$	$2^6 = 64$
$2^2 = 4$	$2^7 = 128$
$2^3 = 8$	$2^8 = 256$
$2^4 = 16$	$2^9 = 512$
$2^5 = 32$	$2^{10} = 1{,}024$

We really do not need the first column because we do not worry about regrouping small collections of data. The second column would be used as shown in the sample problem that follows. Briefly stated we find the closest power of 2 that matches the number of pieces of data in our collection. The exponent becomes the number of intervals we use in our grouped frequency distribution. It should be noted that the text does not use this Guideline, and the number of classes in the exercises may not agree with this approach. Nevertheless, the powers of 2 approach is easy to use when we are asked to determine the number of intervals or classes.

Hot Spot #1 Sample Problem and Solution: Determine the number of intervals for a data collection with 200 pieces of data.

Solution: Step 1.   200 is between 128 and 256.

Step 2.   200 is closer to 256.

Step 3. $256 = 2^8$.

Step 4.   We use 8 intervals.

Sample Problem and Answer: How many intervals should be used for a data collection containing 75 pieces of data?

Answer: 75 is closest to $2^6 = 64$. We use 6 intervals.

Sample Problem and Answer: The data collection contains 500 pieces of data. How many intervals should be used?

Answer: 500 is closest to $2^9 = 512$. We use 9 intervals.

✹
*Johnson &
Bhattacharyya
Section 2.3*

Hot Spot #2 – Frequency Versus Relative Frequency

When we count the number of times we find a particular value in a collection of data, we are getting what is called the frequency of that piece of data. We also find the frequency for each interval in a grouped frequency table by counting how many pieces of data lie in the interval. The relative frequency of either a piece of data or an interval is simply the frequency divided by the number of data in the sample. The formula for relative frequency is as follows:

$$\text{relative frequency} = \frac{\text{Class frequency}}{\text{Total number of observations}}$$

The idea of relative frequency is necessary to create some of the graphs that we use in this chapter. But more importantly, the concept of relative frequency plays an even greater role in what we will call experimental probability in a later chapter. Therefore, relative frequency is an important idea that we need to understand.

If we are given a sample (collection of data) like the following

$$3, 5, 2, 3, 4, 5, 4, 5, 7, 5, 2, 5, 3, 6, 4, 4, 4, 5, 5, 3$$

we can make a frequency table by listing each possible data value and the frequency for each data value as follows:

Data Value	Frequency
2	2
3	4
4	5
5	7
6	1
7	1

We find the relative frequency for each data value by dividing each frequency by $n=20$, the sample size. Note: Be sure to first check that $\sum f = 20$. An easy way to change each frequency to relative frequency is to create a fraction that has the 20 in the denominator as follows:

Data Value	Frequency	Relative Frequency
2	2	$\frac{2}{20}$
3	4	$\frac{4}{20}$
4	5	$\frac{5}{20}$
5	7	$\frac{7}{20}$
6	1	$\frac{1}{20}$
7	1	$\frac{1}{20}$

If we want, we can change each of the fractions to a decimal. When we add all of the relative frequency values together, we should get a sum of 1. Since the sum of the relative frequencies is 1, we can check our work each time by adding the relative frequencies for all of the data values or intervals:

	Relative Frequency
	$\frac{2}{20} = 0.10$
	$\frac{4}{20} = 0.20$
	$\frac{5}{20} = 0.25$
	$\frac{7}{20} = 0.35$
	$\frac{1}{20} = 0.05$
	$\frac{1}{20} = 0.05$
	sum = 1.00 ✔

It is very important when drawing a histogram to check if we are to use frequencies or relative frequencies. When we use frequencies, we get frequency polygons and frequency histograms. In each case the vertical axis gives the frequency for each interval in the grouped frequency table. If the relative frequency is used, the vertical axis gives the relative frequency; the graphs are called relative frequency histograms.

Hot Spot #2 Sample Problem and Solution: Find the relative frequency for each class in Table 2-7a and check that the sum of the relative frequencies totals 1.

City Speeds		
Class	Class Intervals	Frequency
1	20.5–24.5	3
2	24.5–28.5	5
3	28.5–32.5	9
4	32.5–36.5	8

A Study Guide For Statistics

Solution: Step 1. The sample size is 25

Step 2. Change each frequency to a fraction with the
denominator of 25 and write the decimal value

$$\tfrac{3}{25} = 0.12$$

$$\tfrac{5}{25} = 0.20$$

$$\tfrac{9}{25} = 0.36$$

$$\tfrac{8}{25} = 0.32$$

Step 3. Check the sum of the values for equality to 1.0

$$\text{sum} = 1.00 \checkmark$$

Resulting Table:

	City Speeds	
Class	Class Intervals	Relative Frequency
1	20.5–24.5	$\frac{3}{25}$
2	24.5–28.5	$\frac{5}{25}$
3	28.5–32.5	$\frac{9}{25}$
4	32.5–36.5	$\frac{8}{25}$

Sample Problem and Answer: Find both the frequency and relative
frequency for each data value in the following sample:

$$5, 3, 5, 4, 7, 5, 4, 6, 5, 6$$

Answer:

Data Value	Frequency	Relative Frequency
3	1	$\frac{1}{10} = 0.1$
4	2	$\frac{2}{10} = 0.2$
5	4	$\frac{4}{10} = 0.4$
6	2	$\frac{2}{10} = 0.2$
7	1	$\frac{1}{10} = 0.1$

sum = 1.0 ✔

Sample Problem and Answer: Find both the frequency and relative frequency for each class in the following sample:

1.3, 2.6, 1.5, 3.3, 3.8, 1.7, 2.1, 3.2, 4.9, 1.1

Class	Class Intervals	Frequency	Relative Frequency
1	1.05–2.05		
2	2.05–3.05		
3	3.05–4.05		
4	4.05–5.05		

Answer:

Class	Class Boundaries	Frequency	Relative Frequency
1	1.05–2.05	4	$\frac{4}{10} = 0.4$
2	2.05–3.05	2	$\frac{2}{10} = 0.2$
3	3.05–4.05	3	$\frac{3}{10} = 0.3$
4	4.05–5.05	1	$\frac{1}{10} = 0.1$

sum = 1.0 ✔

Hot Spot #3 – Value Of Median Versus Location Of Median.

❋
*Johnson &
Bhattacharyya
Section 2.4*

The median is one of three measures of central tendency that we find in this chapter, and it is easy to find. All we have to do is rank the data and find the middle data value. The formula for the location of the median is also easy to use and gives us the location of the median as

$$\frac{n+1}{2}.$$

When the number of pieces of data is odd, the middle data value is one of the pieces of data. The only real difficulty occurs when the number of data values is even. We then find a value halfway between the two data values in the middle. The two ranked data sets that follow illustrate the difference:

Data Set 1:   3, 3, 4, 7, 9

The location of the median is $\frac{5+1}{2} = 3$.

The median is 4.

Data Set 2:   3, 5, 7, 9

The location of the median is $\frac{4+1}{2} = 2.5$.

The median is the value halfway between 5 and 7.
The median is 6.

Although finding the location of the median and the value of the median is easy to do, an error is often made in confusing the location of the median with the value of the median.

One of the possible explanations for the error is the formula used for the position of the median. The formula is often written as follows:

$$\text{Position} = \frac{n+1}{2}.$$

The mistake may be due to the point that unlike the mean, a formula is not necessary to get the value of the median. It should be clear in the above examples that the value of the median is _not_ the same as the position of the median.

Hot Spot #3 Sample Problem and Solution: Given the following data set, rank the data, find the position of the median, and find the value of the median.

$$4.3, 3.8, 3.6, 4.2, 4.8, 2.1$$

Solution: Step 1. Rank the data: 2.1 3.6 3.8 4.2 4.3 4.8

Step 2. Position $= \dfrac{6+1}{2} = 3.5$

Step 3. The value halfway between 3.8 and 4.2 is 4.0.

Step 4. Median = 4.0

Sample Problem and Answer: Given the following data set, rank the data set, find the location and the value of the median.

$$8.1, 6.0, 5.7, 3.2, 7.1$$

Answer: Step 1. Ranked data: 3.2 5.7 6.0 7.1 8.1

Step 2. Position $= \dfrac{6}{2} = 3$.

Step 3. Median = 6.0

Sample Problem and Answer: Given the following data set, rank the data set, find the location and the value of the median.

$$6, 3, 5, 4, 8, 5$$

Answer: Step 1. Ranked data: 3 4 5 5 6 8

Step 2. Position = 3.5.

Step 3. Median = 5.0

We need to note that 5 is the value halfway between the third and fourth data values; which happen to be the same number in this example.

Hot Spot #4 – $\sum x^2$ Versus $\left(\sum x\right)^2$

Johnson &
Bhattacharyya
Section 2.5

The Greek letter sigma, Σ, is one of the most useful and important notations that we find in statistics. The symbol $\sum$ (sigma) is used to indicate a sum. The expression $\sum x$ means that we are to add up all the data values. Actually, the expression $\sum x$ is a shortened form of the following expression:

$$\sum_{i=1}^{n} x_i = x_1 + x_2 + x_3 + \cdots + x_n$$

The shortened form $\sum x$ is much easier to write and remember. The important point to remember is that whatever follows the sigma sign is to be added together. It may help to think or say "summation of x" rather than "sigma x" when we see $\sum x$.

For example, consider the following data set:

$$2, 8, 4, 9, 5, 6$$

If we want the sum of the data values, we write the expression $\sum x$ and find the summation of x as follows:

$$\sum x = 2 + 8 + 4 + 9 + 5 + 6 = 34$$

If we want to square the sum of the data values, we use a set of parentheses to indicate that the sum is found first, and we write the expression $\left(\sum x\right)^2$. We then square the sum as shown below:

$$\left(\sum x\right)^2 = (34)^2 = 1,156$$

If we want to find the sum of squares, we write the expression $\sum x^2$. We first square each data value. We then add the squares to get our answer:

$$\begin{aligned} \sum x^2 &= 2^2 + 8^2 + 4^2 + 9^2 + 5^2 + 6^2 \\ &= 4 + 64 + 16 + 81 + 25 + 36 \\ &= 226 \end{aligned}$$

It should be very clear that the two summation expressions

$$\left(\sum x\right)^2 \text{ and } \sum x^2$$

do not give the same value. We use both of these expressions in the alternative formula for the variance in this chapter. We will also use the expressions in the next chapter on correlation and regression. It is important to know how to use the sigma notation correctly.

Hot Spot #4 Sample Problem and Solution: Find $\sum x$ and $\sum x^2$ for the following data set: 1, 2, 3, 4, 5

Solution: $\sum x = 1 + 2 + 3 + 4 + 5 = 15$

$\sum x^2 = 1^2 + 2^2 + 3^2 + 4^2 + 5^2 = 1 + 4 + 9 + 16 + 25 = 55$

Sample Problem and Answer: Find $\sum x$ and $\left(\sum x\right)^2$ for the following data set: 1.1, 2.3, 1.2, 2.5, 1.0

Answer: $\sum x = 8.1$

$\left(\sum x\right)^2 = (8.1)^2 = 65.61$

Sample Problem and Answer: Find $\sum x$, $\sum x^2$, and $\left(\sum x\right)^2$ for the following data set: -2, -1, 0, 1, 2

Answer: $\sum x = 0$

$\sum x^2 = 10$

$\left(\sum x\right)^2 = (0)^2 = 0$

*Johnson &
Bhattacharyya
Section 2.5*

Hot Spot #5 – Checking To See That We Have s Rather Than s^2

One of the easiest mistakes to make when calculating the standard deviation is to forget to take the square root. There is an easy way to check on an answer for the standard deviation of distributions that are approximately bell shaped. Generally speaking, the standard deviation should be a value between $\frac{R}{6}$ and $\frac{R}{4}$ where R is the range of the data.

- If the sample size is large, with $n \geq 30$, the standard deviation should be closer to $\frac{R}{6}$. The check value $\frac{R}{6}$ comes from the Empirical Guideline where the distribution is divided into 6 equal regions that are each equal in width to the standard deviation.

- If the sample size is small, with $15 < n < 30$, the standard deviation should be closer to $\frac{R}{4}$. The $\frac{R}{4}$ is an adjustment for small sample sizes between 15 and 30 where we do not often get data from the tails of the curves.

- If the sample size is 15 or less, a check provided by Lincoln E. Moses of Stanford University should be used. The check value for samples smaller than 16 is defined as

$$\frac{R}{\sqrt{n}}$$

and is called the "poor man's standard deviation" (*p.m.s.d.*). Dr. Moses, in his book <u>Think And Explain With Statistics</u>, (Addison-Wesley Publishing Company, 1986), gives credit for the relationship between s and *p.m.s.d.* to Nathan Mantel.

Hot Spot #5 Sample Problem and Solution: Given the following set of data, calculate the sample standard deviation and check the value using one of the check values:

$$2, 7, 5, 13, 10, 4, 15, 2, 10, 2$$

Solution: Step 1. $\bar{x} = \dfrac{\sum x}{n} = \dfrac{70}{10} = 7$

Step 2. $s^2 = \dfrac{\sum (x - \bar{x})^2}{n-1} = \dfrac{206}{9} = 22.89$

Step 3. $s = 4.78$

Step 4. $n = 10 < 16$; use $s \approx \dfrac{R}{\sqrt{n}} = \dfrac{15-2}{\sqrt{10}} = \dfrac{13}{3.16} = 4.11$

Step 5.   The check value agrees with the calculated value of s; note the large difference between the value of s^2 and the check value.

Sample Problem and Answer: The following values were calculated from a data set containing 35 values and having a range of 20. Check the value of the sample standard deviation using one of the check values:

$$n = 35 \qquad\qquad s = 12.2 \qquad\qquad R = 20$$

Answer: Step 1. $n = 35 > 15$; use the $\frac{R}{6}$ and $\frac{R}{4}$ values.

Step 2. $\dfrac{R}{6} = \dfrac{20}{6} = 3.3\overline{3}$ and $\dfrac{R}{4} = \dfrac{20}{4} = 5.0$

Step 3. $s = 12.2$ is not in the interval $\left(3.3\overline{3}, 5.0\right)$. This suggests that the correct value of $s = \sqrt{12.2} = 3.49$.

Organization And Description Of Data

2–17

Sample Problem and Answer: Find the sample standard deviation for the following data set and check the value using one of the check values:

$$2, 3, 7, 3, 5$$

Answer: Step 1. $\bar{x} = \dfrac{20}{5} = 4$

Step 2. $s^2 = \dfrac{16}{4} = 4$

Step 3. $s = 2$

Step 4. Use $\dfrac{R}{\sqrt{n}} = \dfrac{5}{2.236} = 2.236$

Step 5. The check value is comparable to the calculated value of 2: Note the large difference between 2.236 and 4.

Johnson & Bhattacharyya Section 2.5

Hot Spot #6 – The Meaning Of $\bar{x} \pm ks$

The Empirical Guideline requires that we create a symmetric interval about the mean. We use the phrase, $\bar{x} \pm ks$, which reads as the mean plus or minus k standard deviations:

where k is a 1, 2, or 3

For example, if $k = 2$, then we have the interval $(\bar{x} - 2s, \bar{x} + 2s)$. If $\bar{x} = 70$ and $s = 10$ for a distribution of test scores, then the interval for the mean plus or minus two standard deviations would be from 50 to 90.

$$\bar{x} - 2s = 70 - (2)(10) = 70 - 20 = 50$$

and

$$\bar{x} + 2s = 70 + (2)(10) = 70 + 20 = 90$$

If we wanted the mean plus or minus three standard deviations, we would write the expression that follows:

$$\bar{x} \pm 3s$$

$$\bar{x} - 3s \text{ and } \bar{x} + 3s$$

$$\bar{x} - 3s = 70 - (3)(10) = 70 - 30 = 40$$
$$\text{and}$$
$$\bar{x} + 3s = 70 + (3)(10) = 70 + 30 = 100$$

Hot Spot #6 Sample Problem and Solution: Find the interval for the sample mean plus or minus one standard deviation for the above example of test scores. The mean is equal to 70 and the standard deviation is equal to 10.

Solution: $\bar{x} \pm 1s = 70 \pm 10$

$$\bar{x} - 1s = 70 - (1)(10) = 60$$

$$\bar{x} + 1s = 70 + (1)(10) = 80$$

Sample Problem and Answer: Find the interval for $\bar{x} \pm 2.5s$ for the distribution with a sample mean of 1,200 and sample standard deviation of 50.

Answer: The interval goes from 1,075 to 1,325.

Organization And Description Of Data

Sample Problem and Answer: Find the interval for $\bar{x} \pm 3s$ for a distribution with sample mean of 0.5 and sample standard deviation of 0.125.

Answer: The interval goes from 0.125 to 0.875.

*Johnson &
Bhattacharyya
Section 2.4*

Hot Spot #7 – When P_k Does Not Result In $k\%$ Of The Data Below P_k

When we work with the median of a distribution, it is common to argue that half of the data values are below the median and half of the data values are above the median. But as is often the case, we find sets of data that make the above interpretation false. The following data set illustrates the problem:

$$2, 3, 4, 5, 5, 6, 7, 7, 8$$

There are nine data values in the set, and the median is 5. Only three of the data values are lower than the median; whereas, four of the values are above the median. Clearly, in this example, the median of 5 does not separate the data into two sets of equal size. It is because of this type of problem that we define the median as the value that is in the middle of the ranked data list.

A very similar problem occurs when we work with percentiles. The kth percentile, P_k, is supposed to be a value that separates the bottom $k\%$ of the data from the top $(100 - k)\%$ of the data. For example, the median is sometimes defined as P_{50}, the 50*th* percentile. That is the median should separate the set of data into two subsets, each containing 50% of the data.

As we see from the above example, the median of 5 does not separate the data as we would like it to. This is often a problem with small sets of data, but it becomes less of a problem as the data sets get larger. Sometimes we try to get around the problem by saying that the kth percentile is the value at which $k\%$ of the data is at that value or below it. However in the above data set five of the data values are less than or equal to P_{50}. This gives us 56% rather than the desired 50%. Simply stated, there are times when it is impossible to define percentile in a way that avoids inconsistencies.

As with the median, we must be very careful not to confuse the location of the percentile with the value of the percentile. For example, in the above data set the 25*th* percentile is found in the 3*rd* position. We find the position of P_{25} using the formula:

$$\text{Position of } P_k = \left(\frac{kn}{100} \right)$$

To avoid the confusion of position and value, we can get around the problem by writing out the position of P_{25} as follows:

$$\text{Position of the twenty-fifth percentile} = (0.25)(9) = 2.25.$$

We approximate the value of the 2.25 term by averaging the 2nd and 3rd terms in the arranged data as follows:

$$\frac{3+4}{2} = 3.5$$

We reserve the symbol P_{25} for the value of the twenty-fifth percentile as follows:

$$P_{25} = 3.5$$

✳
Johnson &
Bhattacharyya
Sections
2.4 And 2.5

Hot Spot #8 – Rounding Off

One is often tempted to round off the results of calculations like the mean and standard deviation to one or two digits beyond the number of digits after the decimal point in the original data. For example, if the original data values were whole numbers, one usually rounds off the answers to the nearest tenth. This is particularly the case when a calculator is used with a formula to do the calculations for the mean and standard deviation. The practice of rounding off can create difficulties in two ways.

The first problem occurs if one uses the following formula to find the standard deviation:

$$s = \sqrt{\frac{\sum(x - \bar{x})^2}{n - 1}}$$

If the mean is a decimal value that goes beyond two digits after the decimal point, rounding off to the nearest tenth will cause the value of the standard deviation to be incorrect. The error may appear to be small, but the effect on a subsequent value like the standard score is quite significant.

In later chapters we use what is called the standard score or z value to do hypothesis testing. The standard score is defined as follows:

$$z = \frac{\bar{x} - \mu}{\frac{\sigma}{\sqrt{n}}}$$

where the value of s (sample standard deviation) is substituted for σ. Large z values result in a decision to reject the assumption about the population.

If s is in error, the resulting z value will be in error. The decision we then make could also be in error simply because of rounding off.

This then brings up the question of when is it all right to round off. The answer to the question involves two Guidelines:

Rounding Off Guidelines

Guideline I: If we are simply reporting a value such as the mean or standard deviation for a report, then rounding off at the end of each calculation is permissible. As a Guideline of thumb, round off to one more digit after the decimal place than in the original data.

Guideline II: If the value calculated is to be used in a subsequent calculation, the value should not be rounded off. As a Guideline of thumb, one should use as many digits as possible in any calculation that uses the value just obtained. Fortunately calculators allow us to carry as many as 10 digits in any calculations we do. The mean should not be rounded off when it is used to find the standard deviation. The standard deviation should not be rounded off when it is used to find the standard score.

CHAPTER 2 DISCUSSION QUESTIONS

These questions may be used in your study group or simply as topics for individual reflection. Whichever you do, take time to explain verbally each topic to insure your own understanding. Since these questions are intended as topics for discussion, answers to these questions are not provided. If you find that you are not comfortable with either your answers or that your group has difficulty with the topic, take time to meet with your professor to get help.

1. What is the difference between a frequency distribution and a grouped frequency distribution for quantitative data?

2. How do you determine the number of intervals or classes to use with a grouped frequency distribution?

3. What is the difference between frequency and relative frequency?

4. What is the difference between a stem and a leaf?

5. What is a hinge in a box plot?

6. When is it better to use the median rather than the mean?

7. What is meant by the phrase, the mean plus or minus k standard deviations?

8. When is it better to use the alternative formulas for calculating the variance and standard deviation?

9. What is the difference between $\sum x^2$ and $\left(\sum x\right)^2$?

10. What is the difference between the value of the median and the location of the median?

A Study Guide For Statistics

CHAPTER 2 TEST

1. In constructing a grouped frequency distribution, how many intervals or classes should be used for 100 pieces data?

2. If the sample size is 50, what is the relative frequency of the class with a frequency of 12?

3. Draw the relative frequency histogram for the grouped frequency distribution table below:

Class	Class Intervals	Frequency	Relative Frequency
1	14.5–19.5	2	0.2
2	19.5–24.5	7	0.7
3	24.5–29.5	1	0.1

4. Use the following data to construct a stem-and-leaf plot:

$$31 \ 25 \ 34 \ 22 \ 34 \ 46 \ 52 \ 33 \ 49 \ 28 \ 50 \ 37 \ 41 \ 18 \ 62$$

Use the following sample data for problems 5 through 8:

$$4, 2, 3, 5, 2, 8, 9, 4, 5, 8$$

5. Find the position and the value of the median.

6. Find the mean.

7. Find the standard deviation.

8. Find the first quartile.

Use the following information for problems 9 and 10:
$\bar{x} = 70$, $s = 5$; the sample is from a normal population.

9. What percent of the data is in the interval from 60 to 80?

10. What are the values for the interval $\bar{x} = \pm 3s$?

CHAPTER 2 TEST Questions and Answers

1. In constructing a grouped frequency distribution, how many intervals or classes should be used for 100 pieces data?

Answer: 100 is between $2^6 = 64$ and $2^7 = 128$. It is closest to 128, so we use 7 classes.

2. If the sample size is 50, what is the relative frequency of the class with a frequency of 12?

Answer: The relative frequency is $\dfrac{12}{50} = 0.24$

3. Draw the relative frequency histogram for the grouped frequency distribution table below:

Class	Class Intervals	Frequency	Relative Frequency
1	14.5–19.5	2	0.2
2	19.5–24.5	7	0.7
3	24.5–29.5	1	0.1

Answer:

Relative Frequency Histogram

A Study Guide For Statistics

4.   Use the following data to construct a stem-and-leaf display:

31 25 34 22 34 46 52 33 49 28 50 37 41 18 62

Answer:

Stems	Leaves
1	8
2	5, 2, 8
3	1, 4, 4, 3, 7
4	6, 9, 1
5	2, 0
6	2

Stem-and-leaf display

Use the following sample data for problems 5 through 8:

4, 2, 3, 5, 2, 8, 9, 4, 5, 8

5.   Find the position and the value of the median.

Answer: The data, in ranked order, are 2, 2, 3, 4, 4, 5, 5, 8, 8, 9.

$$\text{Position of median} = \frac{n+1}{2} = \frac{11}{2} = 5.5.$$

The median is the value halfway between 4 and 5.

$$\text{Median} = \frac{4+5}{2} = 4.5.$$

6.   Find the mean.

Answer: $\bar{x} = \dfrac{\sum x}{n} = \dfrac{50}{10} = 5$

7. Find the standard deviation.

Answer:

$$s = \sqrt{\frac{n\sum x^2 - \left(\sum x\right)^2}{n(n-1)}} = \sqrt{\frac{10(308) - (50)^2}{9(10)}} = \sqrt{\frac{580}{90}} = \sqrt{6.\overline{4}} = 2.538591$$

8. Find the first quartile.

Answer: The first quartile is $Q_1 = P_{25}$.

The position of P_{25} is $\left(\frac{25}{100}\right)10 = 2.5$. Since 2.5 is not a whole number, we use the average of the *2nd* and *3rd* terms of the data: $\frac{2+3}{2} = 2.5$

Coincidentally, the average equals the position this time.

$$Q_1 = P_{25} = 2.5$$

Use the following information for problems 6 through 8:
$\bar{x} = 70$, $s = 5$; the sample is from a normal population.

9. What percent of the data is in the interval from 60 to 80?

Answer: The interval is $\bar{x} \pm 2s$. Using the Empirical Guideline, 95% of the data is in the interval from 60 to 80.

10. What are the values for the interval $\bar{x} = \pm 3s$?

Answer: $\bar{x} = \pm 3s = 70 \pm 3(5) = 70 \pm 15$

The interval is from 55 to 85.

$$\sum {}^{\circ}$$

chapter three

DESCRIPTIVE STUDY OF BIVARIATE DATA

Choose thou whatever suits the line...
Only, only call me thine.

– Samuel Taylor Coleridge

INTRODUCTION

Our work thus far has involved a single variable. However, many of the problems that we work with in statistics involve two variables. The sets of data for these problems contain ordered pairs. The first value of the ordered pair is called the **independent variable**, and the second value of the ordered pair is called the **dependent variable**. We often refer to our work with two variables as bivariate data analysis. We divide the analysis into two areas, **regression analysis** and **correlation analysis**.

One of the first things we do with bivariate data is plot the ordered pairs on a rectangular coordinate system. In many cases the graph, which is called a scatter diagram, looks like the points are clustered around a straight line. We then say that there appears to be a linear relationship between the two variables. Our first problem is to find the equation of the line that best explains the linear relationship. We use the familiar **slope-intercept form**, $y = mx + b$, of an equation for a line from algebra. The equations we use for m and b involve the Σ notation that we worked with in chapter 2. In particular, we will use the sums, Σx, Σy, Σx^2, Σy^2, and Σxy.

In general, it is impossible to find a line that all the points will fall on. So we focus on the distances (**errors**) between the points and the line we draw. We want the line that minimizes the squares of the distances using the **least squares method**. This line is called the **line of best fit** or the **regression line**. We write the equation of the regression line as $\hat{y} = \hat{\beta}_1 x + \hat{\beta}_0$ where the slope, $m = \hat{\beta}_1$, and the intercept, $b = \hat{\beta}_0$. We are then able to use the equation of this line to predict the approximate value of the dependent variable, $\hat{y}$, when we know the value of the independent variable, x.

If few of the points we plot on the scatter diagram are on the line, we say that the linear relationship is weak with a value of $\hat{\beta}_1$ close to 0. If most of the points of the scatter diagram are on the line or close to the line, we argue that there is a strong linear relationship between the two variables. We then feel good about using the equation for the regression line to predict values for one of the variables.

We use what is called the **correlation coefficient**, r, to measure the strength of the linear relationship. The correlation coefficient is a value between 0 and 1 for a line with a positive slope, and between -1 and 0 for a line with a negative slope. When all the points fall on the line, $r = 1$ or $r = -1$. When there is an absence of a linear relationship, we say that $r = 0$.

A Study Guide For Statistics

CHAPTER 3 HOT SPOTS

1. **How To Write The Equation Of A Line In Slope-Intercept Form.**
   Starts on **page 3–4.** Problems on **pages 3–4, 3–5, 3–6.**
   J&B 3.5

2. **Estimating The Slope And Intercept Of The Regression Line.**
   Starts on **page 3–6.** Problems on **pages 3–9, 3–10.**
   J&B 3.5

3. **The Effect Of Early Rounding Off On The Regression Line And The Correlation Coefficient.**
   Starts on **page 3–11.** J&B 3.4, 3.5

If you find other HOT SPOTS, write them down and use them as a focus of your discussion in the study group. Or you can use the HOT SPOTS as the topic for a help session with your professor.

*Johnson And
Bhattacharyya
Section 3.5*

Hot Spot #1 – How To Write The Equation Of A Line In Slope-Intercept Form

In an algebra class we learn that the equation of a line can be written in the form $y = mx + b$, where m is the slope of the line and b is the y intercept. This form is known as the slope-intercept form for the equation of a line. Once we know the slope–intercept form of the equation, we can use the y intercept and slope of the line to graph the line.

The slope-intercept form we use in regression analysis is very much like going backwards from what we do in an algebra class. In algebra, we change an equation to the slope-intercept form to help us graph the line. In statistics, we have a set of ordered pairs and our problem is to find the equation of the line that best fits the bivariate data. We can use a scattergram to plot the data and draw what we believe to be a good estimate of the line of best fit. We can then use the line to find approximate values for the slope and intercept. Once we have these values, we can write the equation of the line given by our estimates.

In regression analysis, we only have a set of ordered pairs, and our problem is to find the equation of the line that best fits the bivariate data. The next **Hot Spot** shows how we can "eyeball" our best estimate for the line that fits the data. We can then use the line to find approximate values for the slope and intercept. Once we have these values, we can write the equation of the line given by our estimates.

The following example gives an illustration of what we do in an algebra class when we use the slope-intercept form of the equation to graph a line.

Hot Spot #1 Sample Problem and Solution: Given the equation $2x + 3y = 4$, use the slope-intercept form to graph the related line.

Solution: Step 1. We solve for y to get the slope-intercept form as follows:

$$y = -\tfrac{2}{3}x + \tfrac{4}{3}$$

Step 2. The coefficient of x is the slope m= $-\tfrac{2}{3}$. The constant is the y intercept b= $\tfrac{4}{3}$.

Step 3. To graph the line, we first locate the y intercept at $\left(0, \tfrac{4}{3}\right)$ and then use the slope to get a second point by starting at the y intercept and going 3 units to the right, followed by 2 units down. The graph in Figure 3–1 shows how it is done.

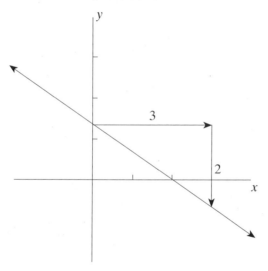

Figure 3–1:
Function Graph
$y = -\tfrac{2}{3}x + \tfrac{4}{3}$

Sample Problem and Answer: Given the equation $3x - 2y = 8$, write the equation in slope-intercept and graph the line.

Answer: $y = \tfrac{3}{2}x - 4$ and $m = \tfrac{3}{2} = 1.5$ and $b = -4$. The graph is shown below, in Figure 3–2.

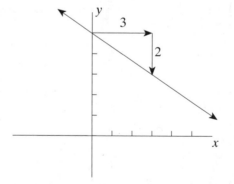

Figure 3–2:
Using The
Slope-Intercept
Form To Graph
3x-2y=8

Sample Problem and Answer: Given the equation $2x + 3y = 15$, write the equation of the line in slope-intercept form and graph the line.

Answer: $y = -\frac{2}{3}x + 5$ and $m = -\frac{2}{3}$ and $b = 5$. The graph is shown in Figure 3–3.

Figure 3–3:
Using The
Slope-Intercept
Form To Graph
2x+3y=15

❋
*Johnson And
Bhattacharyya
Section 3.5*

Hot Spot #2 – Estimating The Slope And Intercept Of The Regression Line

When we have a set of bivariate data, we can plot the points and get what we call a scatter diagram. We then draw a line through the data that to the best of our ability, contains as many points as possible, and minimizes the vertical distance from the points to the line for the points not on the line.

It may take some practice to consistently draw a line that is close to the actual line of least squares. Sometimes it helps to lightly draw several lines through the scatter diagram to get an idea of which one is best. Even then, there may be more than one line that seems acceptable. Just remember, we are only trying to get approximate values for the slope and intercept. If there are two equally reasonable lines, we go ahead and take one at random (perhaps flipping a coin if necessary to choose). The approximate values for the slope and y intercept that we get from the line will be our estimate for the values of the regression line we get using the least squares criterion.

The following scatter diagram (Figure 3–4) will be used as an example. Two lines have been drawn through the bivariate data set. The solid line will be used to estimate the values for the slope and y intercept.

x	y
3	2
3	4
4	3
4	6
5	4
5	6
6	6
6	8
7	7
8	6

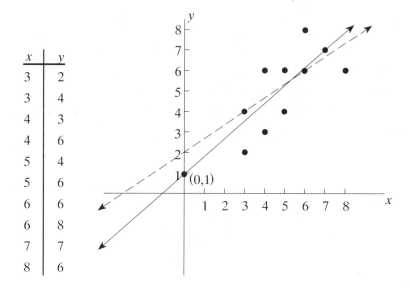

Figure 3-4: Drawing A Regression Line Through A Scatter Diagram

We need to extend the line so it cuts the vertical axis to find the value of the y intercept. In this example the line cuts the vertical axis at 1. We must be careful that both the x and y axis start at the origin without breaks. The modified form of the above graph (Figure 3–4) shows what can happen if we are not careful.

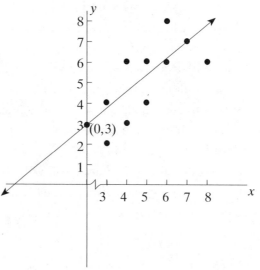

Figure 3–5: The Modified Form Of A Scatter Diagram With Incorrect Y Intercept

The y intercept appears to be positive 3 instead of the correct value of 1. The error is due to the break in the horizontal axis that was made to provide a more pleasing graph.

Returning to Figure 3–4, we find the slope by selecting two points on the line. It helps if the two points are actual ordered pairs from our bivariate data set that are on the line. We then use the slope formula from algebra to find the value of m.

$$m = \frac{y_2 - y_1}{x_2 - x_1}$$

We will use the ordered pairs $(0,1)$ and $(7,7)$ to find the slope.

$$m = \frac{7-1}{7-0} = \frac{6}{7} \doteq 0.857$$

We now have the approximate values for the slope and y intercept. The equation of our estimated line of least squares is

$$y = 0.857x + 1$$

As a comparison, we can get the equation of the line using the principle of least squares. We can use the equations in the text for the slope and intercept of the regression line, or we can use the stat package in the HP–21S. Either way, we get the following equation:

$$\hat{y} = 0.795x + 1.145$$

Our estimate was really quite good. We do not always do this well, but we can at least get an idea of whether or not the values we calculate using the equations for the slope and intercept are reasonable. Even if we use a calculator to get the equation of the regression line, it is important to have a visual sense of what the line should look like before we enter the data into the calculator.

Hot Spot #2 Sample Problem and Solution: Given the following graph of a line, find the slope and y intercept. Then write the equation of the line in the slope-intercept form $y = mx + b$.

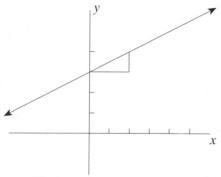

Figure 3–6:
Finding Slope
And Y Intercept
From A Function
Graph

Solution: Step 1. The line crosses the y axis at $(0,3)$.

Step 2. The slope of the line is $\frac{1}{2}$.

Step 3. The equation in slope-intercept form is $y = \frac{1}{2}x + 3$.

Sample Problem and Answer: Given the following graph of a line, find the slope and y intercept. Then write the equation of the line in the slope-intercept form $y = mx + b$.

Figure 3–7:
More Finding Of
Slope And
Y Intercept From
A Function
Graph

Answer: $m = \frac{4}{3}$, $b = (0,-2)$, $y = \frac{4}{3}x - 2$

Sample Problem and Answer: Given the following graph of a line, find the slope and y intercept. Then write the equation of the line in the slope-intercept form $y = mx + b$.

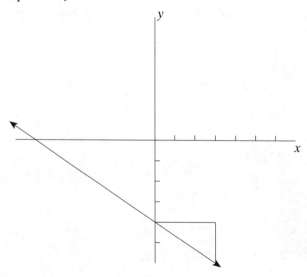

Figure 3–8: Yet
More Finding Of
Slope And
Y Intercept From
A Function
Graph

Answer: $m = -\frac{2}{3}$, $b = (0,-4)$, $y = -\frac{2}{3}x - 4$

Hot Spot #3 – The Effect Of Early Rounding Off On The Regression Line And The Correlation Coefficient

✻

Johnson And Bhattacharyya Sections 3.4 And 3.5

The formulas we use for finding the slope, $\hat{\beta}_1$, and y intercept, $\hat{\beta}_0$, of the line of least squares and the correlation coefficient, r, are very sensitive to errors in the summations we use. This is particularly true if the data values are small. The following set of bivariate data will be used to show what happens when errors are made:

x	y	x^2	y^2	xy
1	1.2	1	1.44	1.2
1.5	0.8	2.25	0.64	1.2
2	0.6	4	0.36	1.2
3	0.4	9	0.16	1.2
4	0.3	16	0.09	1.2

Table 3–1: Table of Sample Bi-Variate Data

The totals for each column are found as follows:

$$\sum x = 11.5 \quad \sum y = 3.3 \quad \sum x^2 = 32.25 \quad \sum y^2 = 2.69 \quad \sum xy = 6$$

When the above sums are substituted into the formulas for $\hat{\beta}_0$, $\hat{\beta}_1$, and r, we get the following values:

$$\hat{\beta}_1 = -0.27413 \qquad \hat{\beta}_0 = 1.2905 \qquad r = -0.9227$$

We are now ready to see what happens if an error is made in one of the entries in the table. One of the most common mistakes is making an error in squaring one of the values in the table. The values of x and y in the above table are easy to square, but let's see what happens when we make an error in squaring $y = 0.3$. If $(0.3)^2$ were written in error as 0.9, the only sum that would change is Σy^2, which would increase to 3.5. In this case the values of $\hat{\beta}_0$ and $\hat{\beta}_1$ do not change because the formulas do not use Σy^2. But there is a drastic change in the r value. The new value of r is -0.574205.

If we look at the value of r^2 for the two different values of r we obtained, we can really see the effect of the error. The r^2 of 0.8514 for the correct value of r tells us that approximately 85% of the variation in the y values is due to the regression line. When the error is made, the new r^2 is approximately 33%, which is a reduction of more than 50% in explained variation.

Now let's look at errors that also affect values of $\hat{\beta}_0$ and $\hat{\beta}_1$. If the error were made in a value of x^2, the error changes the value of the denominator in the formula for both $\hat{\beta}_1$ and r; thus changing the value of all three computations. We can illustrate this last point by looking at what happens if the square of 1.5 is written incorrectly as 22.5 rather than 2.25. The Σx^2 is increased by 20.25 and the effect on $\hat{\beta}_0$, $\hat{\beta}_1$, and r is again drastic. The new values based on the error are as follows:

$$\hat{\beta}_1 = -0.0610 \qquad \hat{\beta}_0 = 0.8004 \qquad r = -0.4354$$

Of the two errors we examined this last one is the more serious because all three values are affected to such a degree that we might as well throw out the entire analysis. Again the point is that care must be taken in the calculations used in the formulas for $\hat{\beta}_0$, $\hat{\beta}_1$, and r.

The obvious solution to this problem of errors is to use a calculator or computer to obtain the values for the regression line and the correlation coefficient. If we must do the calculations using the formulas, it is imperative that we avoid errors in the table of values we create to find $\hat{\beta}_0$, $\hat{\beta}_1$, and r.

CHAPTER 3 DISCUSSION QUESTIONS

These questions may be used in your study group or simply as topics for individual reflection. Whichever you do, take time to explain verbally each topic to insure your own understanding. Since these questions are intended as topics for discussion, answers to these questions are not provided. If you find that you are not comfortable with either your answers or that your group has difficulty with the topic, take time to meet with your professor to get help.

1. How can you use the graph of the bivariate data to determine if the correlation coefficient is positive or negative?

2. How can you use the scatter diagram to get approximate values of the slope and y intercept for the regression equation?

3. What is the major purpose of collecting bivariate data?

4. What is a contingency table?

5. What are the marginal totals?

6. Why is a linear relationship between two variables not necessarily a causal relationship?

7. What is spurious correlation?

8. How does early rounding off of values affect both regression and correlation analysis?

9. Which variable is the predictor variable?, the response variable?

10. How might it be possible that two variables are related even though the linear correlation coefficient is nearly zero?

CHAPTER 3 TEST

Use the following information for problems 1, 2 and 3:

$$\sum x = 10 \qquad \sum y = 14 \qquad \sum xy = 58 \qquad \sum x^2 = 44 \qquad \sum y^2 = 80 \qquad n = 4$$

1. Find $\bar{x}$ and $\bar{y}$.

2. Find $\hat{\beta}_0$ and $\hat{\beta}_1$.

3. Find r.

4. Given the following graph of $y = mx + b$, estimate m and b:

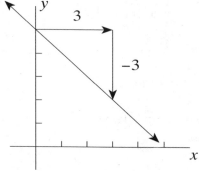

5. Given the regression line $\hat{y} = 1.25x + 4.5$, find a point estimate for y when $x = 1.5$.

6. Given that the ordered pair $(1.5, 6.1)$ belongs to the bivariate data set, find the error for the estimate in problem 5.

7. Plot the scatter diagram of the data:

x	1	3	2	4	3
y	3	5	5	7	6

8. Calculate the correlation coefficient for #7.

9. Compute the least squares estimates $\hat{\beta}_0$ and $\hat{\beta}_1$ for #7.

10. From a visual inspection, draw the line of best fit for #7; and estimate the equation of the line of best fit using the graph.

CHAPTER 3 TEST Questions and Answers

Use the following information for problems 1, 2 and 3:

$$\sum x = 10 \qquad \sum y = 14 \qquad \sum xy = 58 \qquad \sum x^2 = 44 \qquad \sum y^2 = 80 \qquad n = 4$$

1. Find $\bar{x}$ and $\bar{y}$.

Answer: $\bar{x} = \dfrac{\sum x}{n} = \dfrac{10}{4} = 2.5 \qquad \bar{y} = \dfrac{\sum y}{n} = \dfrac{14}{4} = 3.5$

2. Find $\hat{\beta}_0$ and $\hat{\beta}_1$.

Answer: $\hat{\beta}_1 = \dfrac{S_{xy}}{S_{xx}}$

$$\hat{\beta}_1 = \dfrac{(\sum xy) - \dfrac{(\sum x)(\sum y)}{n}}{(\sum x^2) - \dfrac{(\sum x)^2}{n}}$$

$$\hat{\beta}_1 = \dfrac{58 - \dfrac{(10)(14)}{4}}{44 - \dfrac{(10)^2}{4}} = \dfrac{92}{76} = 1.2105$$

$\bar{x} = 2.5 \qquad \bar{y} = 3.5$

$\hat{\beta}_0 = \bar{y} - \hat{\beta}_1 \bar{x} = 3.5 - 3.026 = 0.4737$

3.   Find r.

Answer: $r = \dfrac{SS_{xy}}{\sqrt{SS_{xx}}\sqrt{SS_{yy}}} = \dfrac{\left(\sum xy\right) - \dfrac{\left(\sum x\right)\left(\sum y\right)}{n}}{\sqrt{\left(\sum x^2\right) - \dfrac{\left(\sum x\right)^2}{n}}\sqrt{\left(\sum y^2\right) - \dfrac{\left(\sum y\right)^2}{n}}}$

$$r = \dfrac{(58) - \dfrac{(10)(14)}{4}}{\sqrt{(44) - \dfrac{(10)^2}{4}}\sqrt{(80) - \dfrac{(14)^2}{4}}} = \dfrac{23}{\sqrt{19}\sqrt{31}} = 0.9477$$

4.   Given the following graph of $y = mx + b$, estimate m and b:

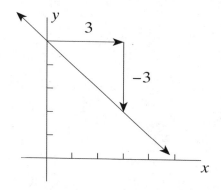

Answer: $m = -1$ and $b = 5$

5.   Given the regression line $\hat{y} = 1.25x + 4.5$, find a point estimate for y when $x = 1.5$.

Answer: $\hat{y} = 1.25(1.5) + 4.5 = 1.875 + 4.5 = 6.375$.

6. Given that the ordered pair $(1.5, 6.1)$ belongs to the bivariate data set, find the error for the estimate in problem 5.

Answer: $\hat{e} = \hat{y} - y = 6.375 - 6.1 = 0.275$.

7. Plot the scatter diagram of the data:

x	1	3	2	4	3
y	3	5	5	7	6

Answer:

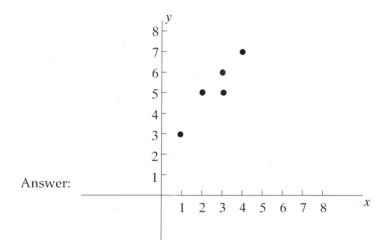

8.   Calculate the correlation coefficient for #7.

Answer: $r = 0.946$

9.   Compute the least squares estimates $\hat{\beta}_0$ and $\hat{\beta}_1$ for #7.

Answer: $\hat{\beta}_1 = 1.23$ and $\hat{\beta}_0 = 2.0$

10.   From a visual inspection, draw the line of best fit for #7; and estimate the equation of the line of best fit using the graph.

Answer:

$$\hat{\beta}_1 = \frac{4}{3} \text{ and } \hat{\beta}_0 = 1.8 \text{ and } \hat{y} = 1.3x + 1.8 \qquad \sum \circ$$

chapter four

PROBABILITY

This branch of mathematics [probability] is the
only one, I believe,
in which good writers frequently get results
entirely erroneous.

— Charles Sanders Pierce

INTRODUCTION

This chapter on probability is often the most challenging chapter in the introductory statistics course. In many respects, one could do a minimal amount of work in probability and still be able to do most of the problems in statistics. This, of course, brings us to the question of "Why do we have this chapter on probability?"

In part, the answer to the above question involves the creation and development of the introductory statistics course in the 1960's. At that time, the usual statistics curriculum was taught in upper division and involved a semester of probability; followed by a second semester of statistics. When the lower division statistics course evolved, a chapter on probability was considered to be an important part of the course.

Aside from the historical perspective, there is good reason to understand the ideas in the probability chapter. Our early work in constructing relative frequency histograms and polygons from the relative frequencies of grouped data provided us with what are called experimental or *a posteriori* probabilities. It only seems natural to extend this idea in a formal way to theoretical or *a priori* probabilities.

Theoretical probabilities are developed using the idea of a **sample space** that results from the outcomes of the experiment. We will focus on the outcomes, called **simple events**. Capital letters, such as A and B are used to represent **events** that are collections or sets of the simple outcomes.

Events are classified in a variety of ways using terms that have an intuitive appeal that helps us understand their meaning. For example, we will talk about events being **independent**, **mutually exclusive**, and **complementary**. We then find the probability of events occurring using the idea of counting. Unfortunately, the counting process can become cumbersome. It is simplified through the use of **Tree Diagrams, Venn Diagrams**, the **Counting Rule** and formulas that focus on the probability of two events occurring. Specifically, we will be interested in the probability of the **compound events** (A and B) and (A or B), and the **conditional probability** of A given B.

We then apply the concepts and formulas by doing problems that involve experiments such as tossing coins, drawing cards, and throwing dice. The problems provide us an opportunity to become familiar with the terms, formulas, and notation used in probability.

Many of the ideas we develop in this chapter are essential to understanding the work we do in the chapters that follow. For example, we use the concept of probability in defining both discrete and continuous probability distributions. These theoretical distributions are used as models for the data distributions we worked with earlier.

We also use the distributions in the hypothesis testing that is such an important part of inferential statistics. In fact, the very essence of hypothesis testing involves finding the probability of getting the value of the sample statistic we use to challenge the accepted value of the population parameter.

CHAPTER 4 HOT SPOTS

If you find other HOT SPOTS, write them down and use them as a focus of your discussions in the study group. Or you can use the HOT SPOT as the topic for a help session with your professor.

USE THIS PAGE FOR KEEPING TRACK OF TERMS AND NOTATION

Hot Spot #1 – Using Tree Diagrams

We will start with the Tree Diagram. It is perhaps the most important topic in this chapter. The Tree Diagram is very similar to a factor tree that we first see in the prime factoring of composite numbers. For example, in factoring 35 we get the following factoring tree:

Johnson And Bhattacharyya Section 4.2

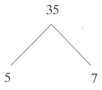

Figure 4–1: 5 × 7 Factor Tree

The factor tree branches out to the two possible factors. We consider the toss of one coin to be like starting with a number. The two outcomes of H (heads) and T (tails) are like the factors. We diagram the outcomes just as we did the factoring:

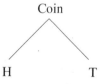

Figure 4–2: Coin Toss Diagram

We usually turn the Tree Diagram on its side as illustrated below:

Tree Diagram

Figure 4–3: Coin Toss Tree Diagram

We can treat the toss of two coins like the factoring of 36. The second factoring resembles the second toss of the coin. The list of factors in the last step is like the list of outcomes we get in the tree diagram. The two diagrams are illustrated below:

Factor Tree

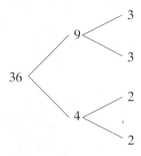

Figure 4–4:
Factor Tree

Tree Diagram

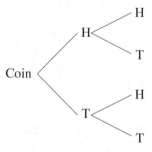

Figure 4–5:
Tree Diagram

We now complete the Tree Diagram by listing the probability of each outcome on the appropriate branch:

Tree Diagram

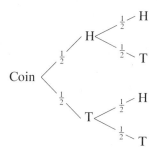

Figure 4–6:
Tree Diagram
w/Probabilities

The Tree Diagram can be used to help us list both the outcomes of an experiment and the probability of each of the outcomes:

Tree Diagram

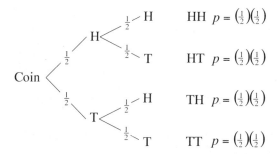

HH $p = \left(\frac{1}{2}\right)\left(\frac{1}{2}\right)$

HT $p = \left(\frac{1}{2}\right)\left(\frac{1}{2}\right)$

TH $p = \left(\frac{1}{2}\right)\left(\frac{1}{2}\right)$

TT $p = \left(\frac{1}{2}\right)\left(\frac{1}{2}\right)$

Figure 4–7:
Tree Diagram
w/Outcomes

There are times when all we do is list outcomes without writing in the probabilities of each outcome. In fact, we may not know the probabilities. We then simply use the Tree Diagram as a tool to help us understand how to do a problem. The following Tree Diagram, without the probabilities written in, illustrates the toss of one die (the singular of "dice"):

Tree Diagram

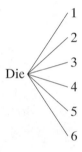

Figure 4–8:
Tree Diagram,
Toss Of One Die

If we were to toss the die a second time the Tree Diagram would be more difficult to draw and require more space. It is drawn on the following page as a solution to the first **Sample Problem and Solution**. The solution will take a full page to draw. This time, neither the probabilities nor the outcomes of the experiment will be listed.

The Tree Diagram is perhaps most useful when we are illustrating probability experiments that involve drawing without replacement. When we include the probabilities for each branch and the list of outcomes for the experiment, the diagram provides us the answers for questions about conditional probability.

As an example, we will draw two cards without replacement from a regulation deck containing 52 cards with four suits and four aces. We are interested in drawing an ace. Each time we draw, we will either get an ace (A) or not an ace ($\overline{A}$). The probability of drawing A on the first draw is $\frac{4}{52}$. But the probability of A on the second draw depends on what happened on the first draw. The probability of $\overline{A}$ on the first draw is $\frac{48}{52}$. But again, the probability of $\overline{A}$ on the second draw depends on what happened on the first draw. The following Tree Diagram helps us organize what is starting to sound complicated:

Tree Diagram

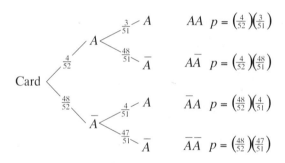

$$AA \quad p = \left(\tfrac{4}{52}\right)\left(\tfrac{3}{51}\right)$$

$$A\overline{A} \quad p = \left(\tfrac{4}{52}\right)\left(\tfrac{48}{51}\right)$$

$$\overline{A}A \quad p = \left(\tfrac{48}{52}\right)\left(\tfrac{4}{51}\right)$$

$$\overline{A}\,\overline{A} \quad p = \left(\tfrac{48}{52}\right)\left(\tfrac{47}{51}\right)$$

Figure 4–9: Tree Diagram, Draw of Two Cards

Hot Spot #1 Sample Problem and Solution: Draw the Tree Diagram for two tosses of a six sided die.

Solution:

Tree Diagram

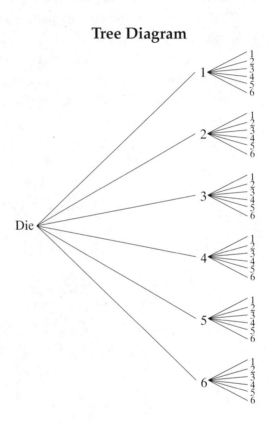

Figure 4–10:
Tree Diagram,
Two Tosses Of A
Die

A Study Guide For Statistics

Sample Problem and Answer: Draw a Tree Diagram for three tosses of a coin. Include the probabilities and the outcomes of the experiment.

Answer:

Tree Diagram

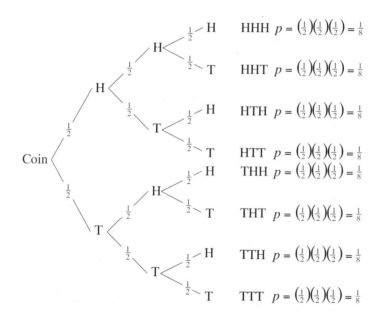

HHH $p = \left(\frac{1}{2}\right)\left(\frac{1}{2}\right)\left(\frac{1}{2}\right) = \frac{1}{8}$

HHT $p = \left(\frac{1}{2}\right)\left(\frac{1}{2}\right)\left(\frac{1}{2}\right) = \frac{1}{8}$

HTH $p = \left(\frac{1}{2}\right)\left(\frac{1}{2}\right)\left(\frac{1}{2}\right) = \frac{1}{8}$

HTT $p = \left(\frac{1}{2}\right)\left(\frac{1}{2}\right)\left(\frac{1}{2}\right) = \frac{1}{8}$

THH $p = \left(\frac{1}{2}\right)\left(\frac{1}{2}\right)\left(\frac{1}{2}\right) = \frac{1}{8}$

THT $p = \left(\frac{1}{2}\right)\left(\frac{1}{2}\right)\left(\frac{1}{2}\right) = \frac{1}{8}$

TTH $p = \left(\frac{1}{2}\right)\left(\frac{1}{2}\right)\left(\frac{1}{2}\right) = \frac{1}{8}$

TTT $p = \left(\frac{1}{2}\right)\left(\frac{1}{2}\right)\left(\frac{1}{2}\right) = \frac{1}{8}$

Figure 4–11:
Tree Diagram,
Three Tosses Of
A Coin

Sample Problem and Answer: Draw the Tree Diagram for the experiment consisting of tossing one coin followed by the draw of one card from a deck where success is drawing an ace. Include the individual probabilities for each branch, the outcomes of the experiment, and their probabilities in the Tree Diagram.

Answer:

Tree Diagram

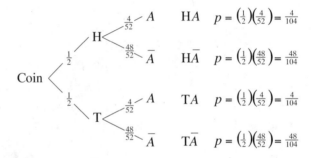

Figure 4–12:
Tree Diagram,
Toss Of One
Coin And Draw
Of One Card

✳
Johnson And
Bhattacharyya
Sections
4.2 And 4.6

Hot Spot #2 – Using The Counting Rule

On the surface, probability seems straight forward. After all, we only need to list the sample space and count which outcomes are a success. Unfortunately, listing outcomes is not always easy; even with a Tree Diagram. For example, the popular dice game, YAHTZEE, involves tossing five dice. This is the same as taking one die and tossing it five times. When we get all five dice with the same number we have a YAHTZEE! It is interesting to find the probability of a YAHTZEE, because it is unreasonable to list the outcomes using a Tree Diagram. Yet, it is easy to find the answer.

What we do is find a way of determining the number of outcomes without counting them. We know that there are only six ways of getting a YAHTZEE. All we need is the number of total possible outcomes to write the correct probability.

In the first **Sample Problem and Solution**, (**Hot Spot #1**) we found that there were 6×6, or 36, outcomes when we used two dice. All we have to do is realize that for every new dice we toss we have six new outcomes. The six new outcomes will be multiplied by the previous six outcomes. When we toss five dice we will have $6 \times 6 \times 6 \times 6 \times 6$, or 7,776 total possible outcomes. The probability of a YAHTZEE is $\frac{6}{7,776}$.

$$P(\text{YAHTZEE}) = \frac{6}{7,776}$$

The principle we just used is called the counting rule. Let us assume we do two things at once where there are k ways of doing the first thing and m ways of doing the second thing. There are then km ways of doing the two things at the same time. If we extend to n ways for doing a third thing, then we would have kmn ways of doing the three things together. The counting rule gives us a way of often finding the number of total outcomes without actually counting or listing the outcomes with a Tree Diagram.

The Counting Rule

Hot Spot #2 Sample Problem and Solution: How many outcomes are there when we toss 5 coins?

Solution: There are two outcomes for each coin. Since we are tossing five coins, we multiply $2 \times 2 \times 2 \times 2 \times 2$ to get 32 outcomes.

Sample Problem and Answer: We toss a six sided die and draw a card from a regulation deck of 52 cards where we record the suit drawn. How many outcomes are possible?

Answer: $6 \times 4 = 24$ outcomes.

Sample Problem and Answer: An experiment involves tossing a coin, tossing a six sided die, and drawing a card from a deck of 52 cards and recording the suit. How many possible outcomes?

Answer: $2 \times 6 \times 4 = 48$ outcomes.

❋
Johnson And Bhattacharyya Section 4.4

Hot Spot #3 – Using The Complement

The YAHTZEE problem we did earlier in **Hot Spot #2** provides us with another insight regarding complementary events. The complement of an event A is everything in the sample space after A is removed. Since there were only six ways of getting a YAHTZEE, there are $7,776 - 6 = 7,770$ outcomes in the complement of the set of six outcomes that gave us a YAHTZEE. With this information, we can find the probability of not getting a YAHTZEE.

$$P(\text{Not a YAHTZEE}) = \frac{7,770}{7,776}$$

A simpler way of finding the probability of "not a YAHTZEE" is to use the probability formula for complementary events:

$$P(\overline{A}) = 1 - P(A)$$

Complementary Events Formula

Using the above formula,

$$P(\overline{Y}) = 1 - P(Y) = 1 - \frac{6}{7,776} = \frac{7,770}{7,776}$$

The probability formula for complementary events is very useful when the problem we are trying to do has many outcomes. The complement of the event then involves few outcomes. Finding the probability of at least one success is one of the classic problems using the idea of a complement. For example, suppose we are tossing a coin three times and we want the probability of at least one tail. We use the complement of at least one tail which is getting zero tails.

$$P(\text{at least one tail}) = 1 - P(\text{zero tails})$$

There is only one way of getting zero tails, namely getting exactly three heads. The answer is easy to set up as follows:

$$P(\text{at least one tail}) = 1 - \left(\tfrac{1}{2}\right)\left(\tfrac{1}{2}\right)\left(\tfrac{1}{2}\right) = 1 - \tfrac{1}{8} = \tfrac{7}{8} = 0.875$$

The key phrase that most often helps us know when to use the complement is "find the probability of at least one success."

Hot Spot #3 Sample Problem and Solution: In a toss of a die three times, what is the probability of getting at least one 3?

Solution: The complement of getting at least one 3 is to get zero 3's. This can be done in only one way. The probability of not getting a 3 is $\frac{5}{6} = 0.83\overline{3}$.

$$P(\text{at least one 3}) = 1 - P(\text{zero 3's})$$
$$= 1 - \left(\tfrac{5}{6}\right)\left(\tfrac{5}{6}\right)\left(\tfrac{5}{6}\right) = 1 - 0.39 = 0.61$$

Sample Problem and Answer: In a production process ten items are selected from a shipment. Assume that each item has a probability of 0.05 of being defective. What is the probability that at least one item will be defective?

Answer: We note that $P(\text{zero defectives}) = 0.95$ for each selection.

$$P(\text{at least one defective}) = 1 - P(\text{zero defectives}) = 1 - (0.95)^{10}$$
$$= 1 - 0.599 = 0.401$$

Sample Problem and Answer: In five tosses of a coin what is the probability of at least two heads?

Answer: The probability of at least two heads tells us we are interested in $x \geq 2$, where x is the number of heads. The complement of $x \geq 2$ is $x \leq 1$. Thus the complement of at least two heads is the event of getting either zero heads or one head. The

$P(\text{zero heads}) = \left(\frac{1}{2}\right)^5$ and the $P(\text{one head}) = 5\left(\frac{1}{2}\right)^5$. If we looked at a Tree Diagram for five tosses of a coin, we would find that five of the thirty-two branches have exactly one head.

$$P(\text{at least two heads}) = 1 - \big(P(\text{zero heads}) + P(\text{one head})\big)$$
$$= 1 - \left(\left(\tfrac{1}{2}\right)^5 + 5\left(\tfrac{1}{2}\right)^5\right) = 1 - \left(\tfrac{1}{32} + \tfrac{5}{32}\right)$$
$$= 1 - \tfrac{6}{32} = 1 - 0.1875 = 0.8125$$

Hot Spot #4 – Independent Events Versus Dependent Events

✳
Johnson And Bhattacharyya Section 4.5

The work with conditional probability leads us to the idea of independent events versus dependent events. The idea of independence suggests that one event does not influence a second event. However, the concept is really defined using a formula irrespective of whether or not the independence appeals to us intuitively. The definition for independent events A and B requires the following to be true:

$$P(A|B) = P(A) \text{ and } P(B|A) = P(B)$$

Independent Events

The two statements are equivalent. For example, If we know $P(A|B) = P(A)$, then we know $P(B|A) = P(B)$. If $P(A|B) \ne P(A)$ or $P(B|A) \ne P(B)$, the events are dependent.

We can illustrate the idea with the following example that involves tossing a six sided die. Find the probability that we get a 3, given we know the outcome is a multiple of 3.

$$\text{Let } A = \{3\} \qquad B = \{3,6\} \qquad S = \{1,2,3,4,5,6\}$$

$$P(A|B) = \frac{P(AB)}{P(B)} = \frac{\frac{1}{6}}{\frac{2}{6}} = \frac{1}{2} \qquad\qquad P(A) = \frac{1}{6}$$

$$P(A|B) \neq P(A)$$

So A and B **are not** independent events; they are dependent events.

In contrast, we will consider a similar problem again using a six sided die. This time we want the probability of an even number, given we know the outcome is a multiple of three.

$$\text{Let } A = \{2,4,6\} \quad B = \{3,6\} \qquad S = \{1,2,3,4,5,6\}$$

$$P(A|B) = \frac{P(AB)}{P(B)} = \frac{\frac{1}{6}}{\frac{2}{6}} = \frac{1}{2} \qquad\qquad P(A) = \frac{1}{2}$$

$$P(A|B) = P(A)$$

So A and B **are** independent events.

We need to note that the last two problems were alike in structure, and the given information was the same. In the first problem, the events were dependent, but in the second problem the events were independent. The point is we must check for independence using the following definition:

Independent
Events Formula

$$P(A|B) = P(A) \underline{\text{ if and only if }} A \text{ and } B \text{ are independent.}$$

Hot Spot #4 – Sample Problem and Solution: Consider the toss of two six sided dice. Let A equal the event *the sum is even* and B equal the event *the sum is a multiple of three.* Determine if A and B are independent.

Solution: There are 36 pairs of outcomes when throwing two dice. (See the tree diagram on page 4–10 of the Study Guide).

$$P(A) = \tfrac{18}{36} \qquad P(AB) = \frac{6}{36} \quad P(B) = \tfrac{12}{36} \qquad P(A|B) = \tfrac{6}{12}$$

Since $P(A) = \tfrac{1}{2}$ and $P(A|B) = \tfrac{1}{2}$, A and B are independent.

Sample Problem and Answer: Given the toss of one six sided die, let A equal the event *the number is more than three* and let B equal the event *the number is a multiple of three.* Determine if A and B are independent.

Answer: $P(A) = \tfrac{3}{6} = \tfrac{1}{2}$ and $P(A|B) = \tfrac{1}{2}$. Therefore, A and B are independent.

Sample Problem and Answer: Consider the toss of a six sided die. Let A equal the event *the number is even* and B equal the event *the number is odd.* Check for independence.

Answer: $P(A) = \tfrac{3}{6}$ $P(A|B) = \tfrac{0}{6}$. Therefore, A and B are dependent.

Johnson And Bhattacharyya Sections 4.4 And 4.5

Hot Spot #5 – Mutually Exclusive or Incompatible Events Versus Independent Events

When A and B cannot both occur together, we say the two events are mutually exclusive or incompatible. Another way of looking at the idea is that the first event excludes the second event. This is illustrated in tossing a die and letting A = odd number and B = even number with $S = \{1,2,3,4,5,6\}$. Events A and B are said to be incompatible. Whenever we have $P(A|B) = 0$, the events are said to be incompatible:

$$P(A|B) = 0 \text{ because } P(AB) = \frac{n(AB)}{n(S)} \text{ and } n(AB) = 0.$$

Mutually Exclusive or Incompatible Events

It is important to understand the difference between events that are incompatible and those that are independent. A simple way of relating the two ideas is that if two events are incompatible they must be dependent; and therefore not independent. For example, consider two events A and B that are incompatible, with $P(A) = 0.5$:

$P(AB) = 0$ because A and B are incompatible.

Using the formula for $P(A|B)$, we get

$$P(A|B) \frac{P(AB)}{P(B)} = \frac{0}{P(B)} = 0$$

Since $P(A) = 0.5$ and $P(A|B) = 0$, the events are dependent.

We must be careful not to get careless in our thinking. It may seem that incompatible events and dependent events are one and the same. This is clearly **not** the case. The following example involves two events that are dependent but not incompatible. The problem of finding the probability of getting a 3 given a multiple of 3 on the toss of one die is done as follows:

Let $A = \{3\}$ $B = \{3,6\}$ with $S = \{1,2,3,4,5,6\}$

$$P(A|B) = \frac{P(AB)}{P(B)} = \frac{\frac{1}{6}}{\frac{2}{6}} = \frac{1}{2} \qquad P(A) = \frac{1}{6}$$

Since $P(A|B) \neq P(A)$, A and B are dependent.

Since $P(AB) \neq 0$, A and B are **not** incompatible.

Hot Spot #6 – Special Vs. General Formula For $P(AB)$ & $P(A \cup B)$

There are two formulas in the chapter that we have not used so far in this review. They are referenced as the addition and the multiplication laws. Actually, we have used the formulas informally as we have worked through some of the problems.

When we considered the probability of the event $A \cup B$, we simply added the probabilities when the events were incompatible. But not all events are incompatible; so we need the addition law that is used when we want the probability of $A \cup B$.

$$P(A \cup B) = P(A) + P(B) - P(AB)$$

In a similar way, when we worked with Tree Diagrams and considered the probability of $A\,B$, we multiplied the probabilities together. If we do not use a Tree Diagram, it may not be clear whether one event is influencing another; as it happened when we drew cards without replacement. As a result, we need what is called the multiplication law that is used when we want the probability of $A\,B$.

$$P(AB) = P(A)P(B|A)$$

or equivalently

Johnson And Bhattacharyya Sections 4.4 And 4.5

P(AUB)
General Case

P(AB)
General Case

$$P(AB) = P(A)P(B|A)$$

In the special cases of independent events and incompatible events, the addition and multiplication laws can be written in much simpler forms.

If events A and B are incompatible events, the probability of $A B$ equals 0. The addition law then becomes the sum of the probabilities:

P(AUB)
Special Case

$$P(A \cup B) = P(A) + P(B)$$

If events A and B are independent, $P(B|A) = P(B)$, then the multiplication law for the probability of AB becomes the product of the probabilities:

P(AB)
Special Case

$$P(AB) = P(A)P(B)$$

Clearly, it is an advantage to be aware of the special cases. But as a rule, it is better to learn the more general formulas that will work for any events, whether or not they are incompatible or independent.

✳
*Johnson And
Bhattacharyya
Sections
4.4 And 4.5*

Hot Spot #7 – Deciding On Which Formula To Use

One of the most difficult things to do in this chapter is to decide on which formula to use. There are a few keys that will help us decide on the correct formula. For example, the use of the phrase "probability of at least one" suggests the complement formula $P(\overline{A}) = 1 - P(A)$. The phrase "determine if events are independent" tells us to use

$$P(A|B) = \frac{P(AB)}{P(B)}$$

Another way to help us decide on the correct formula is to become familiar with the formulas. A good way to do this is to do problems that require we know and correctly use appropriate formulas. The following example illustrates the idea:

Given $P(A) = 0.5$, $P(AB) = 0.25$; find $P(B|A)$.

Step 1.   We need a formula that contains $P(B|A)$ so we use
$$P(AB) = P(A)P(B|A)$$

Step 2.   We substitute the given information into the formula, getting $0.25 = (0.5)P(B|A)$.

Step 3.   We then solve for $P(B|A) = \dfrac{0.25}{0.5} = 0.5$

Hot Spot #7 Sample Problem and Solution: Given $P(AB) = 0.25$, $P(A) = 0.5$, $P(A \cup B) = 0.7$, find $P(B)$.

Solution: Step 1.   We need a formula that includes $P(A)$, $P(AB)$, $P(A \cup B)$ and $P(B)$, so we use

$$P(A \cup B) = P(A) + P(B) - P(AB)$$

Step 2.   We substitute getting $0.7 = 0.5 + P(B) - 0.25$.

Step 3.   We then solve for $P(B) = 0.7 - 0.5 + 0.25 = 0.45$.

Sample Problem and Answer: Given $P(\overline{A}) = 0.3$; find $P(A)$.

Answer: Step 1. We need the formula for the complement
$$P(A) = 1 - P(\overline{A}).$$

Step 2. We substitute, getting $P(A) = 1 - 0.3 = 0.7$.

Sample Problem and Answer: Given $P(A|B) = 0.2$, $P(\overline{B}) = 0.3$, find $P(AB)$.

Answer: Step 1. We need two formulas: $P(B) = 1 - P(\overline{B})$ and
$$P(A|B) = \frac{P(AB)}{P(B)}$$

Step 2. We use the first formula and solve for $P(B)$.
$$P(B) = 1 - 0.3 = 0.7$$

Step 3. We then use the second formula and substitute the value for $P(B)$ and $P(A|B)$.
$$0.2 = \frac{P(AB)}{0.7}$$

Step 4. We solve.
$$P(A \text{ and } B) = (0.2)(0.7) = 0.14$$

Hot Spot #8 – $P(AB)$ Versus $P(A|B)$

❋
Johnson And Bhattacharyya Section 4.5

There is one particularly confusing situation involving the multiplication formula. The formula can be written in two ways as follows:

$$P(AB) = P(A)P(B|A) \text{ or } P(B|A) = \frac{P(AB)}{P(A)}$$

In the first form we are looking at the probability of A and B, and we call it the multiplication law. But in the second form, we are focusing on the probability of A given B. We get the second form by using the first formula and solving for $P(B|A)$. The second form is referred to as the formula for conditional probability.

The confusion occurs when we are asked to find $P(B|A)$, and we realize we need $P(AB)$. It seems like we should be able to use the first formula to find $P(AB)$. But we then realize that this requires $P(B|A)$, which is what we wanted in the beginning. We have just gone full-circle; like a puppy chasing its tail. We are not able to do what we wanted to do. Instead, we must go back to the sample space to find the $P(AB)$ We then substitute our value into the equation

$$P(B|A) = \frac{P(AB)}{P(A)}$$

to get our answer.

Another problem occurs when we realize that there are really two variations of the formula for what we call conditional probability. We can write either

$$P(A|B) = \frac{P(AB)}{P(B)} \text{ or } P(B|A) = \frac{P(AB)}{P(A)}$$

Conditional Probability Formula

The situation is not nearly so confusing if we note the following:

- In the phrase $P(A|B)$, the $A|B$ reminds us of the fraction

$$\frac{A}{B},$$

whose denominator is B.

- On the right side of the equal sign we have

$$\frac{P(AB)}{P(B)}$$

whose denominator involves the same letter B.

We can restate the pattern as follows:

The letter in the denominator on the right side of the equal sign is the same as the letter that follows the vertical bar (|) on the left side of the formula.

Using A Pattern To Find The Denominator In The Formula For Conditional Probability

$$P(B|A) = \frac{P(AB)}{P(A)} \longleftarrow \text{letter in denominator on right side}$$

$\uparrow$

letter after vertical bar

CHAPTER 4 DISCUSSION QUESTIONS

These questions may be used in your study group or simply as topics for individual reflection. Whichever you do, take time to explain verbally each topic to insure your own understanding. Since these questions are intended as topics for discussion, answers to these questions are not provided. If you find that you are not comfortable with either your answers or that your group has difficulty with the topic, take time to meet with your professor to get help.

1. What is the difference between dependent and independent events?

2. Why is the probability of A and B equal to zero when A and B are mutually exclusive?

3. When using a Tree Diagram, why is the number of branches at the end the same as the number of total possible outcomes?

4. What is the difference between marginal probability and conditional probability?

5. What is the difference between the relative frequency concept and the classic definition of probability?

6. When does one use the probability formula $P(A) = 1 - P(\overline{A})$?

7. When is the formula $P(A \cup B) = P(A) + P(B)$ used?

8. When is the formula $P(AB) = P(A)P(B)$ used?

9. When drawing from a container without replacement, what form of the formula for the probability of A and B is used?

10. What rule is used to help decide the number of possible outcomes for a probability experiment?

CHAPTER 4 TEST

1. Draw the Tree Diagram for the experiment of tossing a four sided die (Tetrahedron) where a 1, 2, 3, and 4 are possible. Indicate outcomes only.

2. Fill in the appropriate probabilities on the following Tree Diagram for answering a five part multiple choice test question when guessing.

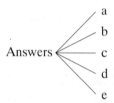

3. How many possible outcomes are there when both a coin and a four sided die are tossed?

4. If $P(\overline{B}) = 0.7$ $P(A|B) = 0.1$, find $P(AB)$.

5. Determine if the following events involving the throw of a six sided die are independent:

    A = getting a multiple of 2 B = getting a 4 or 6

6. Consider the experiment of drawing two cards in succession without replacement; drawing a face card is a success. Find the probability of B = drawing a face card on the second draw given A = a face card was drawn on the first draw.

7. Find the probability of at least one head in four tosses of a fair coin (heads and tails have the same chance of occurring). Document your answer using the appropriate formula.

A Study Guide For Statistics

8.	If events A and B are independent and $P(A) = 0.25$ and $P(B) = 0.75$, find $P(AB)$ and use your answer to find $P(A \cup B)$. Again document your answers by using the appropriate formulas.

9.	Find the $P(A|B)$ where A = multiple of 5 and B = sum greater than 5 (tossing two dice).

10.	Give a counter example for the following statement: If two events are dependent they must be mutually exclusive.

CHAPTER 4 TEST Questions and Answers

1.　　Draw the Tree Diagram for the experiment of tossing a four sided die (Tetrahedron) where a 1, 2, 3, and 4 are possible. Indicate outcomes only.

Answer:

Tree Diagram

Toss 1
2
3
4

2.　　Fill in the appropriate probabilities on the following Tree Diagram for answering a five part multiple choice test question when guessing.

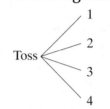

Answers a
b
c
d
e

Answer:

Tree Diagram

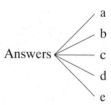

Answers $\frac{1}{5}$ a
$\frac{1}{5}$ b
$\frac{1}{5}$ c
$\frac{1}{5}$ d
$\frac{1}{5}$ e

3.　　How many possible outcomes are there when both a coin and a four sided die are tossed?

Answer:　Total possible outcomes $= 2 \times 4 = 8$

4. If $P(\overline{B}) = 0.7$ $P(A|B) = 0.1$, find $P(AB)$.

Answer: $P(B) = 1 - P(\overline{B}) = 1 - 0.7 = 0.3$

$\qquad P(AB) = P(B)P(A|B) = (0.3)(0.1) = 0.03$.

5. Determine if the following events involving the throw of a six-sided die are independent:

$\qquad A$ = getting a multiple of 2 B = getting a 4 or 6

Answer: $P(A) = \dfrac{3}{6}$ $\qquad\qquad\qquad P(A|B) = \dfrac{P(AB)}{P(B)} = \dfrac{\frac{2}{6}}{\frac{2}{6}} = 1$

$\qquad$ Since $P(A) \neq P(A|B)$, the events A and B are **not** independent.

6. Consider the experiment of drawing two cards in succession without replacement; drawing a face card is a success.
Find the probability of B = drawing a face card on the second draw given A = a face card was drawn on the first draw.

Answer: $P(B|A) = \dfrac{11}{51}$

7. Find the probability of at least one head in four tosses of a fair coin (heads and tails have the same chance of occurring). Document your answer using the appropriate formula.

Answer: $\quad P(\text{at least one head}) = 1 - P(\text{no heads})$

$\qquad\qquad\qquad\qquad\qquad = 1 - \left(\tfrac{1}{2}\right)^4 = 1 - 0.0625 = 0.9375$

8. If events A and B are independent and $P(A) = 0.25$ and $P(B) = 0.75$, find $P(AB)$ and use your answer to find $P(A \cup B)$. Again document your answers by using the appropriate formulas.

Answer: $P(AB) = P(A)P(B) = (0.25)(0.75) = 0.1875$

$$P(A \cup B) = P(A) + P(B) - P(AB)$$
$$= 0.25 + 0.75 - 0.1875 = 0.8125$$

9. Find the $P(A|B)$ where A = multiple of 5 and B = sum greater than 5 (tossing two dice).

Answer: $P(A|B) = \dfrac{P(AB)}{P(B)} = \dfrac{\frac{1}{36}}{\frac{26}{36}} = \dfrac{1}{26}$

10. Give a counter example for the following statement: If two events are dependent they must be mutually exclusive.

Answer: Test question #5 provides the counter example.

A and B are dependent because $P(A) \neq P(A|B)$

A and B are **not** mutually exclusive because $P(AB) \neq 0$

Therefore we have dependent events that are **not** mutually exclusive.

$\sum^{\circ}$

chapter five

PROBABILITY DISTRIBUTIONS

They are called wise
who put things in their right order.

— St. Thomas Aquinas

INTRODUCTION

In this chapter, we bring together many of the ideas developed in the first four chapters. We will again work with the mean and standard deviation and the shapes of distributions. But this time, we will be working with values from the population itself. The populations will be defined using ideas we first saw in the chapter on probability.

We start with the idea of a **random variable** that assigns a number to the outcomes of an experiment. The random variable is classified as either a **continuous random variable** or a **discrete random variable**. The probabilities related to the outcomes are then combined with the random variable to create a **probability distribution**.

Specifically, we will work with **discrete probability distributions** in this chapter. This is the first part of our work with discrete probability distributions that will continue into the next chapter where we study the binomial probability distribution that will be used to do what we will call a hypothesis test in inferential statistics.

In the earlier chapters we looked at collections of sample data. We then calculated sample statistics as estimates of the population parameters, and we were concerned that the samples represented the population without bias. Now we have an opportunity to look at the population itself through the probability distribution. Given a probability distribution, it will now be possible to find the population mean, the population variance, and the population standard deviation. A very useful concept called the **expected value** of a random variable will be defined as the population mean.

The graphs that we created in the earlier chapters provided us an idea about the shape of the distribution of the population. We used the histogram to determine if we had a symmetric bell shaped distribution. In this chapter we create a **probability histogram** using bars to represent the probability for each value of the random variable. The probability histogram, like the grouped frequency distribution histogram, provides insight into the shape of the population defined by the probability distribution.

CHAPTER 5 HOT SPOTS

1. **Discrete Versus Continuous Random Variables.**
   Starts on **page 5–4.** J&B 5.2

2. **Checking For A Probability Distribution.**
   Starts on **page 5–5.** Problems on **pages 5–6.** J&B 5.3

If you find other HOT SPOTS, write them down and use them as a focus of your discussions in the study group. Or you can use the HOT SPOT as the topic for a help session with your professor.

*Johnson &
Bhattacharyya
Section 5.2*

Hot Spot #1 – Discrete Versus Continuous Random Variables

The definitions for both discrete and continuous random variables are rather difficult to understand at first. There is a fairly simple way of thinking about discrete and continuous random variables. If the variable represents something that is counted, then the random variable is discrete. If the variable represents something that is measured, then the random variable is continuous.

It is not enough to just look at the values given for the random variable. For example, if the random variable represents the time to do a statistics problem on a test, we might simply give the time to the nearest five minutes. Even though the time is reported as an integer, the measurement of time could have been any value in an interval of 0 to 60 minutes. Hence, the random variable in this case would be continuous.

It is much easier to determine when we are counting. For example, in this chapter we will count the number of successes in the repeated trials of an experiment. The result is a discrete random variable.

Again, to determine whether the random variable is discrete or continuous, it is best to simply ask if we counted or measured to get the value.

*See the problem
set in section 5.2
for examples of
both discrete
and continuous
variables.*

$$\text{COUNT} \longrightarrow \text{DISCRETE}$$
$$\text{MEASURE} \longrightarrow \text{CONTINUOUS}$$

Hot Spot #2 – Checking For A Probability Distribution

*Johnson &
Bhattacharyya
Section 5.3*

It is easy to check if we have a probability distribution. We must satisfy the following two conditions:

$$0 \le P(x) \le 1 \text{ for all } x$$

$$\sum P(x) = 1$$

The first condition states that the probability of x must be between 0 and 1 for all values of the random variable. The second condition states that the sum of the probabilities for all possible values of the random variable must equal 1. The following example illustrates what should be done to check on a potential probability distribution. Determine if the following function is a probability distribution:

$$P(x) = \frac{x}{5} \text{ for } x = 1, 2, 3, 4, 5$$

We first check to see if $P(x)$ is between 0 and 1 for all x.

$$P(1) = \tfrac{1}{5}$$
$$P(2) = \tfrac{2}{5}$$
$$P(3) = \tfrac{3}{5}$$
$$P(4) = \tfrac{4}{5}$$
$$P(5) = \tfrac{5}{5}$$

The function satisfies the first check. Now we need to find the sum of all the probabilities.

$$P(1) + P(2) + P(3) + P(4) + P(5) = \tfrac{1}{5} + \tfrac{2}{5} + \tfrac{3}{5} + \tfrac{4}{5} + \tfrac{5}{5}$$

$$\sum P(x) = 3$$

The function fails the second check.

Probability Distributions

Hot Spot #2 Sample Problem and Solution: Determine if the following function is a probability distribution function:

$$P(x) = \frac{x^2}{3} \text{ for } x = 1, 2, 3$$

Solution: $P(x) > 1$ for $x = 2$ and $x = 3$.

This is not a probability distribution function.

Sample Problem and Answer: Determine if the following table is a probability distribution:

x	$P(x)$
0	0
1	0.1
2	0.4
3	0.9

Answer: The sum of the probabilities is more than 1.

This is not a probability distribution.

Sample Problem and Answer: Determine if the following is a probability distribution:

x	0	1	2	3
$P(x)$	$\frac{1}{8}$	$\frac{3}{8}$	$\frac{3}{8}$	$\frac{1}{8}$

Answer: Each probability is between 0 and 1, and the sum of the probabilities is 1; therefore, this is a probability distribution.

CHAPTER 5 DISCUSSION QUESTIONS

These questions may be used in your study group or simply as topics for individual reflection. Whichever you do, take time to explain verbally each topic to insure your own understanding. Since these questions are intended as topics for discussion, answers to these questions are not provided. If you find that you are not comfortable with either your answers or that your group has difficulty with the topic, take time to meet with your professor to get help.

1. What is the difference between a discrete random variable and a continuous random variable?

2. What is a random variable?

3. What is a probability distribution?

4. How do you check to see if you have a probability distribution?

5. What is the formula for the mean of a discrete probability distribution?

6. What is the expected value of a random variable?

7. What is the formula for the variance of a discrete probability distribution?

8. What are the two conditions that a probability distribution of a discrete random variable must satisfy?

9. What is a probability histogram?

10. What is the purpose behind drawing a probability histogram?

CHAPTER 5 TEST

Given the discrete probability distribution below:

x	0	1	2	3
$P(x)$	0.1	0.2	0.3	0.4

1. Verify that the table is a discrete probability distribution.

2. Find μ.

3. Find σ^2.

4. Find σ.

Given the following table:

x	0	1	2	3
$P(x)$	0.125	0.375	0.375	0.125

5. Verify that the table represents a discrete probability distribution.

6. Find μ.

7. Find σ^2.

8. Draw the probability histogram.

9. Comment on the pattern of the probability histogram.

10. Find $E(x)$ for the following probability distribution:

x	-2	3
$P(x)$	$\frac{1}{3}$	$\frac{2}{3}$

CHAPTER 5 TEST Questions and Answers

Given the discrete probability distribution below, :

x	0	1	2	3
$P(x)$	0.1	0.2	0.3	0.4

1. Verify that the table is a discrete probability distribution.

Answer: $0 \le P(x_i) \le 1$ and $\sum P(x_i) = 0.1 + 0.2 + 0.3 + 0.4 = 1.0$

2. Find μ.

Answer: $\mu = \sum xP(x) = 0(0.1) + 1(0.2) + 2(0.3) + 3(0.4) = 2$

3. Find σ^2.

Answer: $\sigma^2 = \sum x^2 P(x) - \mu^2 = 0(0.1) + 1(0.2) + 4(0.3) + 9(0.4) - 4 = 1$

4. Find σ.

Answer: $\sigma = 1$

Given the following table:

x	0	1	2	3
$P(x)$	0.125	0.375	0.375	0.125

5. Verify that the table represents a discrete probability distribution.

Answer: $0 \le P(x_i) \le 1$ and $\sum P(x_i) = 0.125 + 0.375 + 0.375 + 0.125 = 1.0$

6. Find μ.

Answer: $\mu = \sum xP(x_i) = (0)0.125 + (1)0.375 + (2)0.375 + (3)0.125 = 1.5$

7. Find σ^2.

Answer: $\sigma^2 = \sum x^2 P(x_i) - \mu^2$

$$= (0)^2 0.125 + (1)^2 0.375 + (2)^2 0.375 + (3)^2 0.125 - 1.5^2$$

$$= 3 - 2.25 = 0.75$$

8. Draw the probability histogram.

Answer:

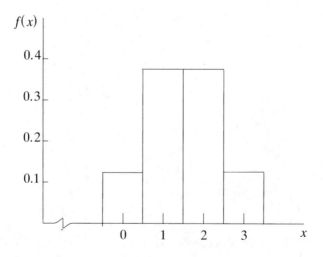

9. Comment on the pattern of the probability histogram.

Answer: The probability histogram is symmetric and somewhat bell shaped in appearance.

10. Find $E(x)$ for the following probability distribution:

x	-2	3
$P(x)$	$\frac{1}{3}$	$\frac{2}{3}$

Answer: $E(x) = \sum xP(x) = (-2)\left(\frac{1}{3}\right) + (3)\left(\frac{2}{3}\right) = -\frac{2}{3} + \frac{6}{3} = \frac{4}{3}$

$\sum\circ$

chapter six

THE BINOMIAL DISTRIBUTION AND ITS APPLICATION IN TESTING HYPOTHESES

Take care of the sense,
and the sounds will take care of themselves.

– Lewis Carroll

INTRODUCTION

In this chapter, we work with one of the most important, and useful, discrete probability distributions. It is named the **binomial distribution** and involves a **binomial experiment** with repeated independent trials that each result in one of two possible outcomes, success or failure. We will use the binomial distribution as a model to find probabilities of many familiar events that we first worked with in the chapter on probability.

This chapter is important to what follows in that we learn how to find the probability of getting specific values of the random variable. In particular, we relate the probabilities to the area in the bars of the probability histogram. We are interested in events that are rare; events that have less than a 5% chance of occurring. When such events do occur, we consider them significant.

The second major part of this chapter extends our work to what is described as **hypothesis testing**. The ideas are simply an extension of what we did in the last two chapters where we found the probability of getting specific values and worked with discrete random variables.

In this chapter the discrete random variable x is from a binomial distribution, and we are given p as a hypothesized probability value someone believes to be true for the binomial distribution. We find the probability that x is larger than or less than some predetermined value that defines a rejection region. If the probability we find is less than 5%, we decide to reject the **hypothesis** and state the percent of the time that we could be wrong.

There are several new terms defined that relate to hypothesis testing. It is important to take the time to learn what the terms mean. An important key to this chapter, and the chapters that follow, is the general structure of hypothesis testing. This process will be used in all of the remaining chapters and involves five parts:

1. Formulate the **null and alternate hypotheses**, H_0 and H_1.
2. State the **test statistic** and the form of the **rejection region**.
3. With a specified value of α, determine the rejection region.
4. Calculate the value of the test statistic.
5. Make a decision to either reject H_o or fail to reject H_o.

The material on hypothesis testing in this chapter is the first part of an introduction to what is called inferential statistics. The ideas on hypothesis testing used with the binomial model in the last section of this chapter are again presented in chapter 9 where we will work with the normal distribution.

Please note that the Hot Spots relating to tests of hypotheses for this chapter are found in chapter 9 to avoid duplicating material. The Hot Spots in this chapter focus on the binomial distribution.

A Study Guide For Statistics

CHAPTER 6 HOT SPOTS

1. **Determining If We Have A Binomial Random Variable.**
   Starts on **page 6–4.** J&B 6.3

2. **Using $P(x) = \binom{n}{x} p^x q^{n-x}$ For Binomial Random Variables.**
   Starts on **page 6–4.** Problems on **pages 6–5, 6–6.** J&B 6.3

3. **Computing $C_{n,x} = \dfrac{n!}{x!(n-x)!}$ Using Pascal's Triangle.**
   Starts on **page 6–6.** Problems on **pages 6–8, 6–9.**
   J&B 6.3

4. **When To Use $\mu = np$ And $\sigma^2 = npq$.**
   Starts on **page 6–10.** J&B 6.3

5. **Interpreting The Probability Of At Least k Successes.**
   Starts on **page 6–11.** Problems on **pages 6–11, 6–12.**
   J&B 6.3

6. **Binomial Probability Distribution Tables**
   Starts on **page 6–13.** Problems on **pages 6–13, 6–14, 6–15.** J&B 6.3

   Please note that the Hot Spots relating to tests of hypotheses for this chapter are found in chapter 9.

If you find other HOT SPOTS, write them down and use them as a focus of your discussions in the study group. Or you can use the HOT SPOT as the topic for a help session with your professor.

Johnson And Bhattacharyya Section 6.3

Hot Spot #1 – Determining If We Have A Binomial Random Variable

Deciding if the random variable is a binomial random variable at first seems like one of the more difficult things to do in this chapter. However, there are only three requirements we need to check.

The first requirement is that the problem involves repeated trials. This is simply a statement that the same thing is done each time. The problems often involve tossing a die repeatedly or flipping a coin repeatedly; but can involve any repetitious task such as doing test questions, shooting baskets, etc.

The second requirement is that each trial is identical and results in one of two possible outcomes. One of the outcomes will be considered a success, the other a failure. The success must be the same in each trial. For example, if we toss a die and decide that getting a 3 is a success, we can not change our mind after 5 tosses and decide that getting a 6 is a success.

The third requirement is that the probability must be the same for each trial. This is not a problem when we toss dice or flip coins. It may indeed be a problem when we shoot baskets or do test questions. In such cases, we may need to make an assumption that the probability of a success does not change from trial to trial.

Johnson And Bhattacharyya Section 6.3

Hot Spot #2 – Using $P(x) = \binom{n}{x} p^x q^{n-x}$ For Binomial Random Variables

There is an important pattern to observe when using the binomial formula to calculate probabilities for a binomial random variable. The exponent x used on p, the probability of success, is the same x in $P(x)$.

A Study Guide For Statistics

For example, if $n = 5$ and $p = 0.2$, then $P(3)$ is written as follows:

$$P(3) = \binom{5}{3}(0.2)^3(0.8)^2$$

We need to notice where the 3's are placed. We have the first 3 in $P(3)$. The second 3 is in $\binom{5}{3}$. The last 3 is in $(0.2)^3$. This pattern is always used when we use the binomial probability formula to find $P(x)$.

Hot Spot #2 Sample Problem and Solution: Given x has a binomial distribution with $n = 4$ and $p = 0.3$, write the expression for $P(3)$.

Solution: Step 1. The 3 in $P(3)$ is the key number. We write $\binom{n}{x}$ as $\binom{4}{3}$

and p^x as p^3.

Step 2. The answer will be in the form $\binom{4}{3}(0.3)^3(0.7)^1$

Sample Problem and Answer: Given the binomial random variable with $n = 8$ and $p = 0.4$, write the expression for $P(5)$.

Answer: The answer will be in the form $\binom{8}{5}(0.4)^5(0.6)^3$

Sample Problem and Answer: Given the binomial distribution with $n = 5$ and $p = \frac{1}{3}$, write the expression for $P(4)$.

Answer: The answer will be in the form $\binom{5}{4}\left(\frac{1}{3}\right)^4\left(\frac{2}{3}\right)^1$

✳

Johnson And
Bhattacharyya
Section 6.3

Hot Spot #3 – Computing $C_{n,x} = \dfrac{n!}{x!(n-x)!}$ Using Pascal's Triangle

The expression $\dfrac{n!}{x!(n-x)!}$ is used in the formula for $P(x)$ for a binomial random variable. Another way of writing the expression is as what we call a combinatorial. We use the notation $C_{n,x}$ to represent $\dfrac{n!}{x!(n-x)!}$. The notation $C_{n,x}$ is used when we want to know how many different combinations are possible when we take n items x at a time.

For example, if we have 5 items such as the letters a, b, c, d and e, and want to take 3 letters at a time, we would write $C_{5,3}$. If we only need to do one computation, the formula using factorials is not that difficult to use. But if we need to do several calculations, as would be necessary in creating a binomial probability distribution table, there is an easier way to find the values of $C_{n,x}$.

The computation can be made very easy if we use Pascal's triangle. The triangle is created by looking at the coefficients of the binomial expansion $(p+q)^n$ for $n = 0, 1, 2, 3, 4, $, etc. The first four expansions are as follows:

$$(p+q)^0 = 1$$
$$(p+q)^1 = 1p^1 + 1q^1$$
$$(p+q)^2 = 1p^2 + 2p^1q^1 + 1q^2$$
$$(p+q)^3 = 1p^3 + 3p^2q^1 + 3p^1q^2 + 1q^3$$

We create Pascal's triangle by writing the coefficients of the terms in the following form:

$$1$$
$$1 \qquad 1$$
$$1 \qquad 2 \qquad 1$$
$$1 \qquad 3 \qquad 3 \qquad 1$$

Pascal's Triangle

We get the next row of the triangle by adding the numbers in the row with 1 3 3 1 as follows:

This new row gives the coefficients of the terms in $(p+q)^4$:

$$(p+q)^4 = 1p^4 + 4p^3q^1 + 6p^2q^2 + 4p^1q^3 + 1q^4$$

We get the next row of the triangle by adding the numbers in the row with 1 4 6 4 1 as follows:

This new row gives the coefficients of the terms in $(p+q)^5$:

$$(p+q)^5 = 1p^5 + 5p^4q^1 + 10p^3q^2 + 10p^2q^3 + 5p^1q^4 + 1q^5$$

We can continue expanding the triangle to as many rows as we need.

It is important to study the pattern of the exponents in the expansions of $(p+q)^n$. We first need to observe that the sum of the exponents for each term is always equal to the value of n. Next, as we examine the terms going from left to right, the exponent on p goes down 1 each time as the exponent on q goes up 1 each time. Again, we

The Binomial Distribution And Its Application In Testing Hypotheses will need to focus on the value of x in the formula $P(x) = C_{n,x}p^x q^{n-x}$.

Now we can use Pascal's Triangle to find $P(x)$ for a binomial random variable. The only difficulty is to decide which term of the triangle to use. All we need to do is be aware of the pattern in the terms of the expansion. If we need to find $P(x)$, the probability of x for a particular problem, we will know n and get x from the notation $P(x)$. The value of n tells us to use the $n+1$ row of the triangle. The value of x gives us the exponent we will use on p for the appropriate term of the expansion. The following problems illustrate the process.

Hot Spot #3 Sample Problem and Solution: Given the binomial distribution with $n = 3$ and $p = 0.1$, find $P(2)$.

Solution: Step 1. $n = 3$, so we want the row of the triangle for $(p+q)^3$.

Step 2. $P(2)$ tells us $x = 2$, so we want the term $C_{3,2}p^2 q^1$.

Step 3. Substituting for p and q we get $P(x) = C_{3,2}(0.1)^2(0.9)^1$.

Step 4. We go to Pascal's triangle to the row 1 3 3 1 and use the first 3 for the value of $C_{3,2}$.

Step 5. We can either leave the answer in the form $P(2) = 3(0.1)^2(0.9)^1$, or perform the computation to get $P(2) = 0.027$.

Sample Problem and Answer: Given the binomial distribution with $n = 4$ and $p = 0.3$, find $P(1)$.

Answer: Step 1. $n = 4$, so we want the row of the triangle for $(p+q)^4$.

Step 2. $P(1)$ tells us $x = 1$, so we want the term $C_{4,1}p^1q^3$.

Step 3. Substituting for p and q we get $P(1) = C_{4,1}(0.3)^1(0.7)^3$.

Step 4. We go to Pascal's triangle to the row 1 4 6 4 1 and use the second 4 for the value of $C_{4,1}$.

Step 5. We can either leave the answer in the form
$P(1) = 4(0.3)^1(0.7)^3$ or perform the computation to get
$P(1) = 0.4116$.

Sample Problem and Answer: Given the binomial random variable with $n = 5$ and $p = \frac{1}{4}$, use Pascal's triangle to find $P(x)$ for $x = 5,4,3,2,1,0$.

Answer:

x	$P(x)$
5	$1\left(\frac{1}{4}\right)^5 = 0.0010$
4	$5\left(\frac{1}{4}\right)^4\left(\frac{3}{4}\right)^1 = 0.0146$
3	$10\left(\frac{1}{4}\right)^3\left(\frac{3}{4}\right)^2 = 0.0879$
2	$10\left(\frac{1}{4}\right)^2\left(\frac{3}{4}\right)^3 = 0.2637$
1	$5\left(\frac{1}{4}\right)^1\left(\frac{3}{4}\right)^4 = 0.3955$
0	$1\left(\frac{3}{4}\right)^5 = 0.2373$

Hot Spot #4 – When To Use $\mu = np$ And $\sigma^2 = npq$

The general formulas for the mean and variance of a discrete probability distribution are:

$$\mu = \sum xP(x) \text{ and } \sigma^2 = \sum (x - \mu)^2 P(x)$$

The formulas will work for any discrete probability distribution, but they are time consuming to use. Fortunately, we can use an easier and faster formula when we have a binomial random variable.

The formulas for the mean and variance of a binomial random variable are:

$$\mu = np \text{ and } \sigma^2 = npq$$

Remember, these shortcut formulas can only be used if the random variable is from a binomial experiment. If the formulas are used for a non-binomial discrete probability distribution, we will get an incorrect answer. The following example shows what happens when we use the wrong formula.

The following function is a probability distribution:

x	1	2	3	4	5	6
$P(x)$	$\frac{1}{6}$	$\frac{1}{6}$	$\frac{1}{6}$	$\frac{1}{6}$	$\frac{1}{6}$	$\frac{1}{6}$

We can use the formula $\mu = \sum xP(x)$ to find the mean because this formula will give us the mean for any discrete probability distribution.

$$\mu = 1\left(\tfrac{1}{6}\right) + 2\left(\tfrac{1}{6}\right) + 3\left(\tfrac{1}{6}\right) + 4\left(\tfrac{1}{6}\right) + 5\left(\tfrac{1}{6}\right) + 6\left(\tfrac{1}{6}\right)$$

$$\mu = (1 + 2 + 3 + 4 + 5 + 6)\left(\tfrac{1}{6}\right) = \tfrac{21}{6} = 3.5$$

Now, we will make the faulty assumption that we have a binomial distribution and use the wrong formula for the problem.

$$\mu = np = 6\left(\tfrac{1}{6}\right) = 1$$

In this problem, it is very clear that the distribution is not a binomial. Furthermore, it is easy to see that the distribution is not centered at $\mu = 1$. Hopefully, we would see our mistake if we used the wrong formula. Of course, it is best to avoid the mistake in the first place and use the correct formula. When in doubt, do not use the shortcut formula.

Hot Spot #5 – Interpreting The Probability Of At Least k Successes

✳
Johnson And Bhattacharyya Section 6.3

The phrase, "find the probability of at least k successes", often causes some difficulty. Part of the problem is to understand what the phrase, *at least*, means. For example, *at least 2* means 2 or more. So if $n = 5$ in a discrete probability distribution, the phrase *at least 2* would mean that we are interested in 2, 3, 4, or 5 successes.

The probability of at least 2 successes would be written as $P(x \geq 2)$ and include $P(x = 2) + P(x = 3) + P(x = 4) + P(x = 5)$. The idea of a complement from the probability chapter could be used to simplify the problem. We would write $P(x \geq 2)$ as $1 - P(x \leq 1)$. Then, we would only have to find $P(x = 0) + P(x = 1)$ and subtract the sum from 1.

Hot Spot #5 Sample Problem and Solution: Given the binomial distribution with $n = 5$ and $p = 0.4$, find the probability of at least one success.

Solution: Step 1. We want to find $P(x \geq 1)$. We use the complement to write the following:

$$P(x \geq 1) = 1 - P(x < 1) = 1 - P(x = 0)$$

Step 2. $P(x = 0) = C_{5,0}(0.4)^0(0.6)^5 = 0.07776$

Step 3. $P(x \geq 1) = 1 - 0.07776 = 0.92224$

Sample Problem and Answer: Given the binomial distribution with $n = 4$ and $p = 0.2$, find the probability of at least 2 successes.

Answer: $P(x \geq 2) = 1 - P(x \leq 1) = 1 - \left[P(x = 1) + P(x = 0)\right]$

$$= 1 - \left[4(0.2)^1(0.8)^3 + 1(0.8)^4\right]$$

$$= 1 - (0.4096 + 0.4096) = 1 - 0.8192 = 0.1808$$

Sample Problem and Answer: Given the binomial distribution with $n = 20$ and $p = 0.5$, find the probability of at least one success.

Answer: $P(x \geq 1) = 1 - P(x = 0) = 1 - (0.5)^{20}$

$$= 1 - 0.000000954 = 0.999999046$$

Hot Spot #6 – Binomial Probability Distribution Tables

✳
*Johnson And
Bhattacharyya
Section 6.3*

When we work with a binomial random variable, there are two common options that are used to find the probability for a specific value of the random variable. The easiest method is to use either a computer or a calculator. When we input a value of n and p, the calculator or computer will give us the probability we want.

The second method is to use a table of probabilities. Binomial probability tables are readily available for small values of n and common values of p, such as $0.1, 0.2, 0.3$, etc. Later on we will work with a method that gives approximate answers for large values of n. However, there may still be a need to construct tables for value of p not found in a typical set of tables. We can then use the binomial probability formula $P(x) = C_{n,x}\left(p^x q^{n-x}\right)$ to construct a table of binomial probabilities. The easiest way to do the problem is using the appropriate expansion of $(p + q)^n$ as shown in the following sample problems.

Hot Spot #6 Sample Problem and Solution: Create the probability table for the binomial probability distribution with $n = 3$ and $p = 0.4$.

Solution: Step 1. We want $(p + q)^3 = 1p^3 + 3p^2 q^1 + 3p^1 q^2 + 1q^3$.

Step 2. We create the table as follows:

x	$P(x)$
3	$1p^3$
2	$3p^2 q^1$
1	$3p^1 q^2$
0	$1q^3$

*See Hot Spot #3
for help with
using Pascal's
Triangle.*

Step 3. We now substitute for p and q as follows:

x	$P(x)$
3	$1(0.4)^3$
2	$3(0.4)^2(0.6)^1$
1	$3(0.4)^1(0.6)^2$
0	$1(0.6)^3$

Step 4. We can write the values with decimals as follows:

x	$P(x)$
3	0.064
2	0.288
1	0.432
0	0.216

Sample Problem and Answer: Given the binomial random variable with $n = 2$ and $p = 0.8$, create the binomial probability table.

Answer:

x	$P(x)$
2	$1(0.8)^2$
1	$2(0.8)^1(0.2)^1$
0	$1(0.2)^2$

Sample Problem and Answer: Given the binomial distribution with $n = 5$ and $p = \frac{1}{3}$, create the table of binomial probabilities.

Answer:

x	$P(x)$
5	$1\left(\frac{1}{3}\right)^5$
4	$5\left(\frac{1}{3}\right)^4\left(\frac{2}{3}\right)^1$
3	$10\left(\frac{1}{3}\right)^3\left(\frac{2}{3}\right)^2$
2	$10\left(\frac{1}{3}\right)^2\left(\frac{2}{3}\right)^3$
1	$5\left(\frac{1}{3}\right)^1\left(\frac{2}{3}\right)^4$
0	$1\left(\frac{2}{3}\right)^5$

*See **Hot Spot #3** for help with using Pascal's Triangle.*

CHAPTER 6 DISCUSSION QUESTIONS

These questions may be used in your study group or simply as topics for individual reflection. Whichever you do, take time to explain verbally each topic to insure your own understanding. Since these questions are intended as topics for discussion, answers to these questions are not provided. If you find that you are not comfortable with either your answers or that your group has difficulty with the topic, take time to meet with your professor to get help.

1. What is a Bernoulli Trial?

2. What is a binomial random variable?

3. What is a binomial probability distribution?

4. What is meant by the probability of at least one?

5. How is Pascal's Triangle used to find the coefficients of $(p+q)^n$?

6. When can you use $\mu = np$ and $\sigma^2 = npq$?

7. What is a statistical hypothesis?

8. What is the difference between a null and alternate hypothesis?

9. What are the five steps of testing a hypothesis?

10. Describe the two types of errors that can be made when testing a hypothesis and give the notation for the probability of each error.

A Study Guide For Statistics

CHAPTER 6 TEST

1. Write out the probability distribution table for the binomial random variable with $n = 2$ and $p = \frac{1}{2}$.

2. Given the binomial distribution with $n = 5$ and $p = \frac{1}{2}$, find μ.

3. Given the binomial distribution with $n = 5$ and $p = \frac{1}{2}$, find σ^2.

4. Find the value of $\dfrac{n!}{x!(n-x)!}$ for $n = 5$ and $x = 3$.

5. Given the binomial random variable with $n = 5$ and $p = 0.2$ use the formula $P(x) = C_{n,x}p^x q^{(n-x)}$ to find $P(3)$.

6. Given the binomial distribution with $n = 3$ and $p = 0.6$. Find the probability of at least one success.

    In a test of $H_0 : p \le 0.6$ versus $H_1 : p > 0.6$ with $n = 10$, the decision rule is to reject H_0 if $x \ge 8$.

7. Determine the type of error that can occur and the error probability when $p = 0.9$.

8. Determine the type of error that can occur and the error probability when $p = 0.4$.

9. Redo #7 and #8 when $n = 20$ and the rejection region is $x \ge 16$.

10. What should the rejection region be if the desired level of significance is approximately 5%?

CHAPTER 6 TEST Questions and Answers

1. Write out the probability distribution table for the binomial random variable with $n = 2$ and $p = \frac{1}{2}$.

Answer:

x	$P(x)$
0	$\frac{1}{4}$
1	$\frac{1}{2}$
2	$\frac{1}{4}$

2. Given the binomial distribution with $n = 5$ and $p = \frac{1}{2}$, find μ.

Answer: $\mu = np = (5)\left(\frac{1}{2}\right) = 2.5$

3. Given the binomial distribution with $n = 5$ and $p = \frac{1}{2}$, find σ^2.

Answer: $\sigma^2 = npq = 5\left(\frac{1}{2}\right)\left(\frac{1}{2}\right) = \frac{5}{4}$

4. Find the value of $\dfrac{n!}{x!(n-x)!}$ for $n = 5$ and $x = 3$.

Answer: $\dfrac{5!}{(3!)(2!)} = \dfrac{(5)(4)(3)(2)(1)}{[(3)(2)(1)][(2)(1)]} = \dfrac{(5)(4)(3)(2)(1)}{(3)(2)(1)(2)(1)} = \dfrac{20}{2} = 10$.

5. Given the binomial random variable with $n = 5$ and $p = 0.2$ use the formula $P(x) = C_{n,x} p^x q^{(n-x)}$ to find $P(3)$.

Answer: $P(3) = C_{5,3}(0.2)^3(0.8)^2 = 10(0.008)(0.64) = 0.0512$.

6. Given the binomial distribution with $n = 3$ and $p = 0.6$. Find the probability of at least one success.

Answer: $p(x \geq 1) = 1 - p(x = 0) = 1 - (0.4)^3 = 1 - 0.064 = 0.936$

In a test of $H_0: p \leq 0.6$ versus $H_1: p > 0.6$ with $n = 10$, the decision rule is to reject H_0 if $x \geq 8$.

7. Determine the type of error that can occur and the error probability when $p = 0.9$.

Answer: Type II error and when $p = 0.9$, $\beta = p(x < 8) = 0.07$

8. Determine the type of error that can occur and the error probability when $p = 0.4$.

Answer: Type I error and when $p = 0.4$,
$$\alpha = p = (x \geq 8) = 1 - p(x \leq 7) = 1 - 0.988 = 0.012$$

9. Redo #7 and #8 when $n = 20$ and the rejection region is $x \geq 16$.

Answer: $\alpha = p(x \geq 16) = 1 - p(x \leq 15) = 1 - 0.9997 = 0.0003$
$\beta = p(x < 16) = p(x \leq 15) = 0.04317$

10. What should the rejection region be if $\alpha \leq 0.05$ when $n = 20$?

Answer: $p(x \geq 17) = 1 - p(x \leq 16) = 1 - 0.984 = 0.016$

$\sum \circ$

chapter seven

THE NORMAL DISTRIBUTION

Everybody believes in the [normal approximation],
the experimenters because
they think it is a mathematical theorem,
the mathematicians because
they think it is an experimental fact.

– G. Lippmann

INTRODUCTION

Many of the ideas in this chapter on **continuous probability distributions** are extensions of material we have already seen in the chapter on discrete probability distributions. The idea of probability as area will again be used. But the area will be under a curve rather than in the rectangles of the probability histograms we used in the last chapter.

There are several key differences in this chapter. In very simple terms, a discrete variable meant that we were counting something. In this chapter, a **continuous variable** is the result of measuring something rather than counting. There is a major difference in the probability statements we will make using discrete and continuous variables. In a discrete probability distribution, it was possible to find the probability of getting a specific value of the random variable. When we work with a continuous random variable, the probability of the variable being equal to a specific value is zero. But we are able to find the probability that a variable is in an interval. The **Empirical Guideline** will be used to find the probability that the continuous variable is in an interval about the mean.

This chapter provides an important transition between the idea of hypothesis testing and a probability distribution; first introduced in the last chapter and then extended to continuous variables in this chapter.

Throughout this chapter, we will use the z score. We will refer to z as the **standardized variable**. The z score and the bell shaped **normal distribution** become the focal points in this chapter. In various forms, the z score will continue to be a focus in the next several chapters. We will also use the z score in the form of a test statistic to help us organize the material in the chapters that follow.

In this chapter, notation continues to be very important to us as a way of organizing our work and understanding what it is that we have done. We will also do a great deal of drawing; it is important that we draw a picture of a bell-shaped curve for each problem that we do.

The last thing we will look at in this chapter is the **normal approximation** to the binomial distribution. Recent advances in calculators and the use of statistical software on personal computers make this section less important than it would have been in the past decade. But the approximation will help us better understand the relationship between the discrete and continuous random variable.

CHAPTER 7 HOT SPOTS

If you find other HOT SPOTS, write them down and use them as a focus of your discussions in the study group. Or you can use the HOT SPOT as the topic for a help session with your professor.

✳

*Johnson And
Bhattacharyya
Section 7.1*

Hot Spot #1 – A Difference Between The Probability For Discrete And Continuous Random Variables

In this chapter we work with many of the same ideas we developed and used in the chapter on discrete probability distributions. However, there is one very important difference beyond the over simplified distinction that we use a discrete random variable when we count and a continuous random variable when we measure. In the work we did with discrete probability distributions, we were able to find the probability that a random variable equals a specific value. But, when we find the probability of a continuous random variable having a specific value, we get zero. (This is discussed further in **Hot Spot #2**).

For example, if the random variable x has a binomial distribution with $n = 5$ and $p = 0.5$, we can find $P(x = 3)$. We find the answer by substituting the values of n and p into the binomial probability function

$$P(x) = \frac{n!}{x!(n-x)!}p^x(1-p)^{n-x}$$

In this example, we get

$$P(x = 3) = 10\left(\tfrac{1}{2}\right)^3\left(\tfrac{1}{2}\right)^2 = 0.3125$$
$$\text{where } 10 = \frac{5!}{3!2!}$$

If x is a continuous random variable, the probability that x equals a specific value is zero. Using notation, we would write $P(x = c) = 0$, where c is any real number.

When we work with a continuous random variable, we are <u>only</u> able to find the probability that x is in an interval. For example, we are able to ask for the probability that x is between a and b. The notation we use involves the expression $P(a \le x \le b)$.

We then find the area between a and b under the curve defined by the continuous probability distribution.

Hot Spot #2 – Understanding Why The Probability Of A Continuous Random Variable Equalling A Specific Value Is Zero.

✳
Johnson And Bhattacharyya Section 7.1

It is sometimes difficult to understand why the probability of a specific value is zero. One way of looking at the problem is to consider a large paper bag in which we write the ten digits between 0 and 9 on slips of paper and place them in the bag. If we then draw out one value, the probability of getting one of the digits between 0 and 9 would be 0.1. Now if we consider adding slips of paper to represent the "tenths" place we could find the probability of getting a specific value to the tenths place between 0 and 9.9. For example we could find the probability of getting a value like 3.4, but now we would have one hundred slips of paper in the bag. The probability of getting one of the values in the bag is now 0.01.

We could then add more slips of paper to the bag for the "hundredths" place so that we can find the probability of getting a value like 3.44. But the probability of getting a specific value is now down to 0.001. As we continue adding slips of paper to the bag to allow us to get the probability of getting any real number between 0 and 10, we would get smaller and smaller probabilities that would get closer and closer to zero. As you realize, we would never be done adding slips of paper to our bag; and we would finally simply say that the probability of getting a specific real number in the half-open interval of 0 to 10 is zero.

Hot Spot #3 – Finding Probabilities Under $y = ax + b$

✳
Johnson And Bhattacharyya Section 7.1

One of the simplest examples of a probability distribution that we can look at for a continuous random variable is given by the probability function $f(x) = 2x$ for $0 \leq x \leq 1$.

The probability function is a linear function. The graph of the equation $y = 2x$ in the interval $[0,1]$ is drawn below:

Figure 7–1:
Graph of
Function y=2x.

We can check to see if the area under the triangle is 1; as it should be for a continuous probability distribution. The height of the triangle at $x = 1$ is found by taking $f(1) = 2$.

$$\text{Area} = \frac{bh}{2} = \frac{(1)(2)}{2} = 1$$

Even though we cannot find the probability that x equals a specific value in the interval from 0 to 1, we can find the probability that the continuous random variable is between two values. We do this by finding the area under the line $y = 2x$ between the two values given in the problem.

For example, we can find the probability that x is between 0.25 and 0.5. We state the problem as follows:

$$\text{Find } P(0.25 \le x \le 0.5)$$

We first look at the graph we drew and locate the two values of 0.25 and 0.5 on the x axis. We then draw a vertical line through the two points on the axis as is shown in Figure 7–2, with the area between the two vertical lines shaded in.

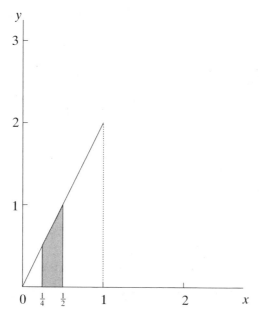

Figure 7–2:
Graph of
Probability Area

We now find the answer to our question by looking at the area of two triangles. The area of the larger triangle is the probability that x is between 0 and 0.5 as shown in Figure 7–3.
We use $A = \dfrac{bh}{2}$.

Figure 7–3:
Graph of Larger
Triangle

$$P(0 \le x \le 0.5) = (0.5)(0.5)(1) = \tfrac{1}{4}$$

The area of the smaller triangle is the probability that x is between 0 and 0.25 as shown below. Again we use $A = \dfrac{bh}{2}$.

Figure 7–4:
Graph of Smaller
Triangle

$$P(0 \le x \le 0.25) = (0.5)(0.25)(0.5) = 0.0625$$

The difference of the two areas is the answer to our original question of finding the probability that x is between 0.25 and 0.5. Using Δ to represent a triangle, the answer to the problem is

$$
\begin{aligned}
P(0.25 \le x \le 0.5) \quad &= \text{Area of large } \Delta - \text{Area of small } \Delta \\
&= P(0 \le x \le 0.5) - P(0 \le x \le 0.25) \\
&= 0.25 - 0.0625 = 0.1875
\end{aligned}
$$

A Study Guide For Statistics

$$P(0.25 \le x \le 0.5) = 0.1875$$

It is important that we use the above notation. It helps us see that the $P(0.25 \le x \le 0.5)$ is the difference between the areas under the line. This last idea is used with all the probability functions we will see in the following chapters.

Hot Spot #3 Sample Problem and Solution: Given the continuous probability function

$$f(x) = \frac{2}{3} - \frac{2x}{9} \text{ for } 0 \le x \le 3$$

find $P(1 \le x \le 2)$; that is, we want the probability that x is in the closed interval $[1,2]$.

Solution: Step 1. We first draw the line given by $y = \frac{2}{3} - \frac{2x}{9}$ in the interval $[0,3]$.

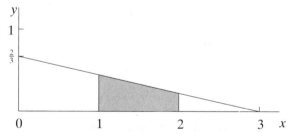

Figure 7–5:
Sample Interval
Graph

Step 2. Check that the area in the triangle is 1.

Area $= \left(\frac{1}{2}\right)(3)\left(\frac{2}{3}\right) = 1$, so $P(0 \le x \le 3) = 1$.

Step 3. The area of the large triangle between 1 and 3 is

$\left(\frac{1}{2}\right)(2)\left(\frac{4}{9}\right) = \frac{4}{9}$.

Step 4. The area of the small triangle between 2 and 3 is

$$\left(\tfrac{1}{2}\right)(1)\left(\tfrac{2}{9}\right) = \tfrac{1}{9}.$$

Step 5. Again using Area of large Δ – Area of small Δ, the

$$P(1 \le x \le 2) = P(1 \le x \le 3) - P(2 \le x \le 3) = \tfrac{4}{9} - \tfrac{1}{9} = \tfrac{1}{3}.$$

Sample Problem and Answer: Given the continuous probability function $f(x) = \tfrac{1}{8}x$, in the interval $0 \le x \le 4$, find the $P(0 \le x \le 2)$.

Answer: $P(0 \le x \le 2) = \tfrac{1}{4}$.

Sample Problem and Answer: Given the continuous probability function $f(x) = \tfrac{1}{3} - \tfrac{x}{18}$, in the interval $0 \le x \le 6$, find the probability that x is between 2 and 4 inclusively.
Hint: $P(2 \le x \le 4) = P(2 \le x \le 6) - P(4 \le x \le 6)$.

Answer: $P(2 \le x \le 4) = \tfrac{4}{9} - \tfrac{1}{9} = \tfrac{1}{3}$.

✹
Johnson And
Bhattacharyya
Section 7.2

Hot Spot #4 – Using The Empirical Guideline to Find $P(a \le x \le b)$

We are usually interested in finding the probability that a continuous random variable is within some interval. For example, let x equal the speed of a car on a highway in miles per hour. If x is assumed to be normally distributed with $\mu = 55$ and $\sigma = 5$, then we could find the probability that the speed of the car is more than 65.

A Study Guide For Statistics

There are two ways for us to look at this problem: using the Empirical Guideline, or using a z score; which is the topic of the next **Hot Spot**. In this **Hot Spot**, we will use the Empirical Guideline. Although this will be easy to do, we must realize our answer will only be an approximation of the correct answer we obtain using the z score.

We know that 95% of the speeds will be within two standard deviations of the mean. The mean is 55 m.p.h. and two standard deviations gives us 10 m.p.h.. When we use 55 for the mean, and then add and subtract the 10, we get the values of 45 and 65. So, 95% of the speeds are in the interval $(45,65)$.

This leaves approximately 5% of the area under the bell-shaped curve outside the interval from 45 to 65. The areas outside of the interval are called the tails of the curve. The area under the right tail of the curve gives the probability of $x \geq 65$. The area under the right tail of the curve is one-half of 5%, and we can write:

$$P(x \geq 65) = 2.5\%$$

Hot Spot #4 Sample Problem and Solution: Use the Empirical Guideline to find the approximate answer for $P(50 \leq x \leq 60)$ where the continuous random variable x equals the speed on the highway with $\mu = 55$ and $\sigma = 5$.

Solution: The interval between 50 and 60 is centered on 55. The graph for the bell-shaped curve centered on 55 follows (Figure 7–6):

The Normal Distribution

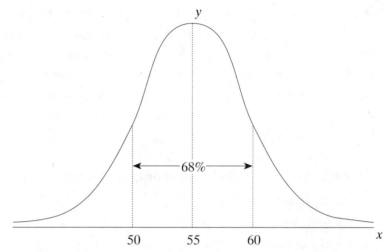

Figure 7–6:
One Standard
Deviation From
Mean of 55

The values of 50 and 60 represent the mean plus and minus one standard deviation. We know that approximately 68% of the distribution will be within one standard deviation of the mean. We can now answer the question using the appropriate notation:

$$P(50 \leq x \leq 60) = 0.68$$

Sample Problem and Answer: Let x be a continuous random variable with a normal distribution with $\mu = 1,200$ and $\sigma = 100$. Use the Empirical Guideline to find an approximate value for $P(x \geq 1,300)$.

Answer: $P(x \geq 1,300) = 0.16$.

Sample Problem and Answer: Let x represent the scores on a statistics test given to all the statistics students at a university. Assume that x has a normal distribution with $\mu = 70$ and $\sigma = 10$. Use the Empirical Guideline to find the approximate probability for $P(60 \le x \le 90)$.

Answer: $P(60 \le x \le 90) = 0.34 + 0.475 = 0.815$.

Hot Spot #5 – Using The Z Score To Find $P(x)$

Johnson And
Bhattacharyya
Sections 7.3, 7.4

An alternative to using the Empirical Guideline to find $P(x)$ is to trade in questions about x for questions about z. The z score compares a data value's relationship to the mean in terms of the standard deviation. We do this so that we can obtain more accurate answers and make it easier to get the answers. Rather than having to figure out in our head how many standard deviations 65 m.p.h. is from 55 m.p.h. in our example about highway speeds with $\mu = 55$ and $\sigma = 5$, we just use the formula

$$z = \frac{x - \mu}{\sigma}$$

We substitute 65 for x and use the values of μ and σ given:

$$z = \frac{65 - 55}{5} = 2$$

The $z = 2$ tells us that we are two standard deviations away from the mean. Now we can replace our original question about the

$$P(x \ge 65) \text{ with } P(z \ge 2)$$

Remember, it does not matter if we use $\ge$ or just $>$ because the

$$P(x = 65) = 0 \text{ and } P(z = 2) = 0$$

Furthermore, it does not matter whether we use the x value or the z value. Either way, we want the area under the appropriate bell-shaped curve as shown in Figures 7–7 and 7–8:

Figure 7–7:
Figure 7–8:
Graphs Of x And
z Distributions
Compared

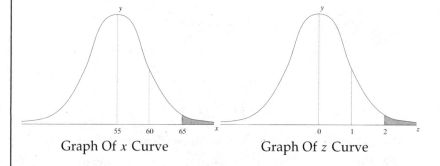

Graph Of x Curve   Graph Of z Curve

The real problem is to find the area under the x curve to the right of 65. There are several methods that can be used, and some methods would take a great deal of time or require a computer to get an accurate answer. For example, one of the methods we could use involves a numerical technique of integration that is done in calculus.

Fortunately, the z curve gives us an answer to the problem very quickly. All we have to do is use the z table in the back of the book. We look up the value of 2 and find that the probability of z being between 0 and 2 is 0.4772 (to the nearest ten thousandth).

Since 50% of the area under the curve is to the right of $z = 0$, we subtract our value of 0.4772 as follows:

$$P(z \geq 2) = 0.5 - 0.4772 = 0.0228$$

Figure 7–9, below, shows the bell-shaped curve for z and the area under the curve between 0 and 2. As is illustrated, we subtract the area that is shaded from 0.5 to get the area to the right of 2.

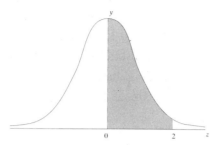

Figure 7–9: Area
Under The Curve
From 0 to 2

Hot Spot #5 Sample Problem and Solution: Use the z value to find the exact answer for $P(50 \leq x \leq 60)$; where the continuous random variable x equals the speed on the highway with $\mu = 55$ m.p.h. and $\sigma = 5$ m.p.h..

*Compare the
next three
problems to the
corresponding
problems of
Hot Spot #4*

Solution: Step 1.   Graph the bell-shaped curve centered on 55 m.p.h..

*Figure 7–10:
Bell Curve
Centered At 55*

Step 2.   Place the values of 50 and 60 on the graph.

Step 3.   We want the area under the curve between the two
values. We now find the z values for each value of x.

$$z = \frac{50 - 55}{5} = -1 \qquad\qquad z = \frac{60 - 55}{5} = 1$$

Step 4.   We now use the table of z values to find the area
between 0 and 1.

$$P(0 \leq z \leq 1) = 0.3413$$

The Normal Distribution

7–15

Step 5. Because the area between 0 and 1 is the same as the area under the curve between -1 and 0, we double the value we found in the z table. We can now answer the question using the appropriate notation:

$$P(50 \leq x \leq 60) = 2 \cdot P(0 \leq z \leq 1) = (2)(0.3413) = 0.6826$$

Sample Problem and Answer: Let x be a continuous random variable with a normal distribution with $\mu = 1,200$ and $\sigma = 100$. Use the z score to find the exact answer for $P(x \geq 1,300)$.

Answer: $P(x \geq 1,300) = 0.1587$

Sample Problem and Answer: Let x represent the scores on a statistics test given to all the statistics students at a university. Assume that x has a normal distribution with $\mu = 70$ and $\sigma = 10$. Use the z score to find the exact probability for $P(60 \leq x \leq 90)$.

Answer: $P(60 \leq x \leq 90) = 0.3413 + 0.4772 = 0.8185$

✳
Johnson And
Bhattacharyya
Sections 7.3, 7.4

Hot Spot #6 – Using Graphs With The Z Score

There is a very important point that needs to be made about working with z values. **Always take time to draw a graph!** The picture we draw helps us see the areas we need to use to get an answer. For example, if we use the normal distribution for x, where x represents highway speeds with $\mu = 55$ and $\sigma = 10$, we can find the probability that x is between 60 and 70.

We would write this using the following notation:

$$P(60 \leq x \leq 70)$$

The first thing we want to do is draw a diagram or graph:

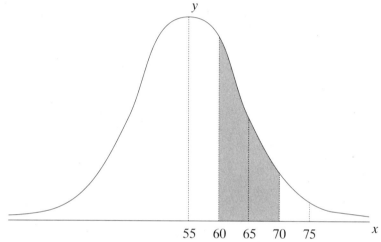

Figure 7–11:
Probability Of x
Between
60 and 70

Since 60 and 70 are both larger than the population mean of 55, the 60 and 70 are placed on the right side of the curve as shown. We want the area under the curve between 60 and 70. Using the type of approach we followed earlier, we could try to use the Empirical Guideline. But this time, we find that we need the area under the bell-shaped curve between one-half and one and one-half standard deviations; $60 = 55 + 0.5(10)$ and $70 = 55 + 1.5(10)$. The Empirical Guideline does not help in this situation. Unless we want to use a technique such as numerical integration from calculus, we have no choice but to use the z values that correspond to 60 and 70 as follows:

$$z = \frac{60 - 55}{10} = \frac{1}{2} = 0.5 \qquad z = \frac{70 - 55}{10} = \frac{3}{2} = 1.5$$

The graph we drew now helps us decide how to find the answer to our original question. Our question is now written as

$$P(0.5 \leq z \leq 1.5)$$

If we decide to use the z table, we will get two different areas as shown below:

Figures 7–12 And 7–13: Two Different z Table Areas

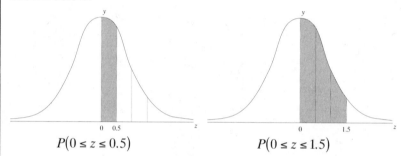

$$P(0 \leq z \leq 0.5) \qquad\qquad P(0 \leq z \leq 1.5)$$

The larger area is the one between 0 and 1.5; we need to subtract the smaller area between 0 and 0.5 from the larger area to get our answer. We find the areas from the z table. Using the table, we look up the 1.5 value for z and find the large area is 0.4332; it is $P(0 \leq z \leq 1.5)$. When we look up 0.5 in the table, we find the small area is 0.1915; it is $P(0 \leq z \leq 0.5)$. Now, all we have to do is use the graph to show us what to do. We need to find the difference of the two areas. Even if we somehow did the subtraction backwards, we would realize our mistake because **the probability must be a positive number**. Using the values we obtained from the table we subtract as follows:

$$P(0.5 \leq z \leq 1.5) = P(0 \leq z \leq 1.5) - P(0 \leq z \leq 0.5)$$
$$= 0.4332 - 0.1915$$
$$P(0.5 \leq z \leq 1.5) = 0.2417$$

It is important to document how we get the answer using both the graph and probability notation. Many students try to take shortcuts at this point by either leaving out the notation or not drawing the graph.

Although it is possible to get the correct answer, students often get confused and make errors when they use shortcuts. So, be careful; document your work with a graph and probability notation.

Hot Spot #6 Sample Problem and Solution: Find $P(-2 \leq z \leq -0.5)$.

Solution: Step 1. Draw the bell-shaped curve.

Step 2. Locate -2 and -0.5 on the curve and shade in the area under the curve between the values.

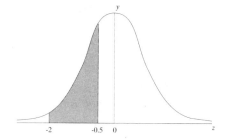

Figure 7–14:
Area Between
-2 and -0.5

Step 3. We see that we need to subtract two areas:

$$P(-2 \leq z \leq 0) - P(-0.5 \leq z \leq 0)$$

Step 4. Since the z table only uses positive values we actually find the following areas:

$$P(0 \leq z \leq 2) - P(0 \leq z \leq 0.5) \quad = 0.4772 - 0.1915$$
$$= 0.2857$$

Step 5. We should rewrite our answer as follows:

$$P(-2 \leq z \leq -0.5) \qquad = P(-2 \leq z \leq 0) - P(-0.5 \leq z \leq 0)$$
$$= 0.4772 - 0.1915 = 0.2857$$

$$P(-2 \leq z \leq -0.5) = 0.2857$$

The Normal Distribution

7–19

Sample Problem and Answer: Find $P(-1.5 \leq z \leq -0.5)$.

Answer: $\begin{aligned} P(-1.5 \leq z \leq -0.5) \quad &= P(-1.5 \leq z \leq 0) - P(-0.5 \leq z \leq 0) \\ &= 0.4332 - 0.1915 \\ &= 0.2417 \end{aligned}$

$$P(-1.5 \leq z \leq -0.5) = 0.2417$$

Sample Problem and Answer: Find $P(z \geq -1.5)$.

Answer: $\begin{aligned} P(z \geq -1.5) \quad &= P(-1.5 \leq z \leq 0) + 0.5 \\ &= 0.4332 + 0.5 \\ &= 0.9332 \end{aligned}$

$$P(z \geq -1.5) = 0.9332$$

✳
Johnson And
Bhattacharyya
Section 7.3

Hot Spot #7 – Finding c For $P(z \geq c) = 5\%$

Related to the problem of finding the probability of z being between two values is the inverse problem of finding the value of z that gives a specific area to the right of z. For example, we may want to find the value of z that gives 5% to the right of our value. This problem can be stated in two different ways. The easiest way is to use the notation $z(0.05)$. There is a more interesting way of asking the questions as follows:

Find c so that $P(z \geq c) = 0.05$

The letter c is used to represent the actual value of z that makes the statement true. In our example we know that c must be on the right side of the curve because the area we want to the right of c is less than

50%. If we now draw the graph with c on the right side, we get the following:

Figure 7–15:
Use Of c With z
Scores

As is usually the case, it is very important to first draw the graph and locate c on the correct side. It is now easier to see that we need a value of c that gives 0.45 between the center of our curve and the value of c. We can use the z table to find that $c = 1.645$ gives us the 0.45 we need. We now replace the letter c with the 1.645 and write the following:

$$P(z \geq 1.645) = 0.05$$

Of all the concepts in this chapter on continuous probability distributions, this one idea seems to be very difficult to understand. Some of the difficulty seems to be with the algebra of the statement

$$z \geq c.$$

If we consider the above expression to be a problem with inequalities much like what we did in an algebra class, it may make it easier to understand what it means to have $z \geq c$. In the algebra class, we would just graph the inequality on a number line and the solution includes all of the values to the right of c as follows:

Figure 7–16:
$z \geq c$ As An
Inequality On A
Number Line

Hot Spot #7 Sample Problem and Solution: Find the value of c that makes the following statement true: $P(-c \leq z \leq c) = 0.95$.

Solution: Step 1.   We need to locate the values of c and $-c$ on the bell-shaped curve. It is easier this time because we have an interval. The graph is drawn below:

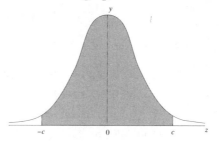

Figure 7–17:
Area Between
-c and c

Step 2.   Since our values of c and $-c$ are symmetric to the center of the curve, we see that we need a value of c that will give us 0.475 between $z = 0$ and our value c. We now look up 0.475 as an area in the z table. We find a value of $c = 1.96$ gives us the required area.

Step 3.   We can now write the following:
$$P(-1.96 \leq z \leq 1.96) = 0.95$$

Step 4.   Our solution is $c = 1.96$.

Sample Problem and Answer: Find the value of c that makes the following statement true: $P(z \leq c) = 0.01$.

Answer: $c = -2.326$

A Study Guide For Statistics

Sample Problem and Answer: Find the z value for $z(0.10)$.

Answer: $c = 1.282$

Hot Spot #8 – Normal Approximation To The Binomial Distribution

❋
Johnson And Bhattacharyya Section 7.5

The last consideration in this chapter on continuous probability distributions involves what is called the normal approximation to the binomial distribution. The normal approximation is very useful if we do not have either a calculator or a computer that is programmed with the binomial distribution. When we have to rely on a binomial table to find probabilities, this approximation is used when either the number of trials is larger than a value listed in a table of binomial probabilities or when the p value is a value not listed in the table. The usual difficulty is illustrated in the following example.

Suppose we are asked to find the probability of getting a score of 70 or more on a 100 question True/False test and that we must assume that the individual taking the test is guessing on every question. If we let x = the number correct, then x is a discrete random variable with $n = 100$ and $p = 0.5$; and we want the $P(x \geq 70)$.

The partial probability histogram below (Figure 7–22) shows the areas we want between 70 and 100; the area includes the rectangles at $x = 70$ and $x = 100$. Whenever np and nq are both greater than or equal to 5, the histogram will be approximately bell-shaped. In our example $np = nq = (100)(0.5) = 50$; so we can use the normal approximation. A bell-shaped curve has been drawn over the probability histogram. The area under the curve will be approximately equal to the area in the rectangles, and we will use the normal curve to approximate the areas in the probability histogram. The curve will be centered at $\mu = np$, the mean of the binomial distribution.

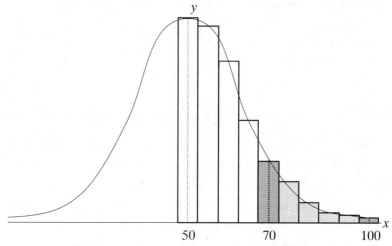

Figure 7–18:
Continuity
Correction
Factor

The area we want includes the rectangles at $x = 70$ and $x = 100$. The rectangle at 70 includes values on the x-axis from 69.5 to 70.5, and the rectangle at 100 includes the values on the x-axis from 99.5 to 100.5. If x is a continuous variable, we want the area under the curve to the right of 69.5. and to the left of 100.5. This is often a confusing point; the important thing to understand is that the following two statements are referencing the same areas:

If x is a discrete variable, we want $70 \leq x \leq 100$.
If x is a continuous variable, we want $69.5 \leq x \leq 100.5$.

Although it looks like we are talking about the same variable we are not. In $70 \leq x \leq 100$, the variable is discrete and it must be a positive integer. In $69.5 \leq x \leq 100.5$, the variable is continuous; there is no longer a restriction that it be an integer. To find the approximate answer to our problem using the normal curve we find

$$P(69.5 \leq x \leq 100.5) \text{ where } \mu = np = 50 \text{ and } \sigma = \sqrt{npq} = 5$$

We now must find the values of the related z scores:

A Study Guide For Statistics

$$z = \frac{69.5 - 50}{5} = 3.9 \qquad z = \frac{100.5 - 50}{5} = 10.1$$

We want the area under the normal curve between 3.9 and 10.1; however, the area to the right of $z = 10.1$ is so small that we can ignore it in this case. We will simply use the area to the right of 3.9; we look up 3.9 in the z table and subtract the value from 0.5 to get our answer:

$P(x \geq 70)$ for discrete variable $\quad = P(x \geq 69.5)$ for continuous variable

$$= 0.5 - 0.49995$$

$$= 0.00005$$

Clearly, if we have either a calculator like the HP-21S or a computer available to get a correct answer, we will not need the normal approximation. But the normal approximation to the binomial distribution provides us with a good opportunity to better understand both distributions. It is because of this opportunity to better understand statistics that we need to spend the time on the approximation even when we have the calculators and computers available.

One more point needs to be addressed regarding the normal approximation. If we had been asked to find the probability of getting more than 70 correct, we could again ignore the area to the right of 100 and write the expression

$$P\left(x_{discrete} > 70\right) = P\left(x_{continuous} \geq 70.5\right)$$

The rectangles for $x > 70$ include rectangles to the right of 70.5 because the first discrete value larger than 70 is 71. It includes the rectangle from 70.5 to 71.5.

Whenever we have a question as to what value to use in the approximation, it helps to draw the partial probability histogram and label it carefully to see which rectangles we need to use. As we can see from the diagram below, it is clear that we need to use 70.5 for the x value of the continuous variable.

Figure 7–19:
Normal
Approximation

68.5   69.5   70.5   71.5   72.5   73.5   etc. x
              70

Hot Spot #8 Sample Problem and Solution: Given a 50 question three-part multiple choice test, what is the probability of getting less than 25 correct when guessing? (Include the area to the left of $x = 0$).

Solution: Step 1. $n = 50$ $p = \frac{1}{3}$ and $q = 1 - p = \frac{2}{3}$

Step 2. $np = \frac{50}{3} = 16.\overline{6}$ and $nq = \frac{100}{3} = 33.\overline{3}$
It is OK to use the normal approximation.

Step 3. $\mu = np \approx 16.7$ $\sigma = \sqrt{npq} = \sqrt{\frac{100}{9}} = \frac{10}{3} \approx 3.3$

Step 4.   We want $P(x < 25) = P(x \le 24)$

Step 5.   Draw a partial graph of the histogram and curve to determine the value of the continuous variable x (See Figure 6–24).

Step 6.   We want $P(x \le 24.5)$ and we need to find the related value of $z = \frac{24.5-16.7}{3.3} = 2.36$.

Step 7. $P(z \le 2.36)$ $= 0.5 + P(0 \le z \le 2.36)$
$= 0.5 + 0.4949 = 0.9909$

Step 8. $P(x < 25) = P(z \le 2.36) = 0.9909$.

A Study Guide For Statistics

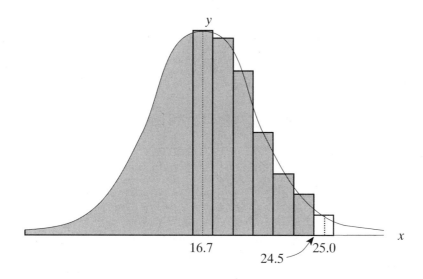

Figure 7–20: Probability Of Getting Less Than 25 Correct When Guessing

16.7   24.5   25.0

Sample Problem and Answer: Given a binomial random variable with 60 trials and $p = \frac{1}{6}$ and $q = 1 - p = \frac{5}{6}$, find the probability that x is greater than 20. (Include the area to the right of 60).

Answer: $P(x > 20) = P(z \geq 3.64) = 0.0001$

Sample Problem and Answer: Given a 40 question True/False test, find the probability of getting between 20 and 30 (inclusive) questions correct when guessing on each question. Hint: Find the z value for $x = 19.5$ and $x = 30.5$.

Answer: $\mu = 40(0.5) = 20$ $\sigma = \sqrt{40(0.5)(0.5)} = 3.16228$

$z = \frac{19.5 - 20}{3.16228} = -0.16$ $z = \frac{30.5 - 20}{3.16228} = 3.32$

$P(-0.16 \leq z \leq 3.32)$ $= P(-0.16 \leq z \leq 0) + P(0 \leq z \leq 3.32)$

$= 0.0636 + 0.4995$

$= 0.5631$

The Normal Distribution

CHAPTER 7 DISCUSSION QUESTIONS

These questions may be used in your study group or simply as topics for individual reflection. Whichever you do, take time to explain verbally each topic to insure your own understanding. Since these questions are intended as topics for discussion, answers to these questions are not provided. If you find that you are not comfortable with either your answers or that your group has difficulty with the topic, take time to meet with your professor to get help.

1. What is the difference between a discrete and continuous random variable?

2. Why is the probability of a specific value of a continuous random variable equal to zero?

3. When using a normal approximation of a binomial random variable what is the difference between the statements $P(x > 4)$ and $P(x \geq 4.5)$?

4. What is the standard normal distribution?

5. What is the relationship of the Empirical Guideline to a z score?

6. When finding c so that $P(z \geq c) = 0.05$, how do you decide which side of the bell-shaped curve to place the letter c?

7. Why is it important to graph the curve and locate the values on the curve when finding $P(1.5 \leq z \leq 2.5)$?

8. Given the continuous random variable x with $\mu = 55$ and $\sigma = 10$ why do we need a z score when finding $P(x \geq 57.5)$?

9. What is the continuity correction for the normal approximation to the binomial?

10. What is the probability density function for the normal distribution?

CHAPTER 7 TEST

1. Use the normal approximation to find the probability of more than 35 successes when $n = 40$ and $p = 0.75$.

2. Given that x has a normal distribution with $\mu = 85$ and $\sigma = 5.25$, find the probability that x is less than 75.

3. Given that x is a continuous random variable with $\mu = 20.5$, $\sigma = 1.25$, and a sample of $n = 49$; find the probability that x is less than 20.

4. Find c so that $P(z \geq c) = 90\%$.

5. Find $P(-1.25 \leq z \leq 0.5)$.

6. Use the calculator to find $P(z \geq 10)$.
    (Write your answer in scientific notation).

7. Given the continuous probability distribution defined by $f(x) = \frac{1}{2}x$ for the interval $[0,2]$, find $P(0.25 \leq x \leq 0.75)$.

8. A sample of 100 measurements of temperature are taken from a population whose value of μ is assumed to be $98.6°$. If the sample standard deviation $s = 1.5$, find $P(x \geq 99)$.

9. If a continuous random variable x has a rectangular (uniform) distribution on the interval $[0,100]$, what is the height of the rectangle?

10. Use the Empirical Guideline to find the probability that test scores will be between 1,150 and 1,300 on a test whose scores are normally distributed with $\mu = 1,200$ and $\sigma = 50$.

CHAPTER 7 TEST Questions and Answers

1.   Use the normal approximation to find the probability of more than 35 successes when $n = 40$ and $p = 0.75$.

Answer: $\mu = np = (40)(0.75) = 30,$

$\sigma = \sqrt{npq} = \sqrt{(40)(0.75)(0.25)} = \sqrt{7.5} = 2.74$. Use $x = 35.5$,

$z = \frac{35.5 - 30}{2.74} = 2.01$. $P(x > 35) = P(z \geq 2.01) = 0.5 - 0.4778 = 0.0222$.

2.   Given that x has a normal distribution with $\mu = 85$ and $\sigma = 5.25$, find the probability that x is less than 75.

Answer: Use $z = \frac{x - \mu}{\sigma}$. $z = \frac{75 - 85}{5.25} = -1.9$.

$P(x < 75) = P(z < -1.9) = 0.5 - 0.4713 = 0.0287$

3.   Given that x is a continuous random variable with $\mu = 20.5$, $\sigma = 1.25$, and a sample of $n = 49$; find the probability that x is less than 20.

Answer: Use $z = \frac{x - \mu}{\sigma}$. $z = \frac{20 - 20.5}{1.25} = \frac{-0.5}{1.25} = -0.4$

$P(x < 20) = P(z < -0.4) = 0.5 - 0.1554 = 0.3446$

4.   Find c so that $P(z \geq c) = 90\%$.

Answer: c must be on the left side of the curve. This leaves 40% between c and 0. The z value to the left of 0 that gives 0.4000 is -1.282; So $c = -1.282$.

5. Find $P(-1.25 \le z \le 0.5)$.

Answer: $P(-1.25 \le z \le 0.5)$ $= P(-1.25 \le z \le 0) + P(0 \le z \le 0.5)$

                                  $= 0.3944 + 0.1915$

         $P(-1.25 \le z \le 0.5)$ $= 0.5859.$

6. Use the calculator to find $P(z \ge 10)$. (Write your answer in scientific notation).

Answer: $7.6198530 \times 10^{-24}$.

7. Given the continuous probability distribution defined by $f(x) = \frac{1}{2}x$ for the interval $[0,2]$, find $P(0.25 \le x \le 0.75)$.

Answer: $A = (0.5)(2)\left(\frac{2}{2}\right) = 1.$ $A_T = (0.5)(0.75)\left(\frac{0.75}{2}\right) = 0.140625.$

         $A_I = (0.5)(0.25)\left(\frac{0.25}{2}\right) = 0.015625.$

         $P(0.25 \le x \le 0.75) = 0.140625 - 0.015625 = 0.125.$

8. A sample of 100 measurements of temperature are taken from a population whose value of μ is assumed to be $98.6°$. If the sample standard deviation $s = 1.5$, find $P(x \ge 99)$.

Answer: Use $z = \dfrac{x - \mu}{s}.$ $z = \dfrac{99 - 98.6}{1.5} = 0.267.$

         $P(x \ge 99) = P(z \ge 0.267) = 0.5 - 0.1053 = 0.3947$

9. If a continuous random variable x has a rectangular (uniform) distribution on the interval $[0,100]$, what is the height of the rectangle?

Answer: The area of the rectangle must be 1. Since the base is 100, the height of the rectangle must be 0.01.

10. Use the Empirical Guideline to find the probability that test scores will be between 1,150 and 1,300 on a test whose scores are normally distributed with $\mu = 1,200$ and $\sigma = 50$.

Answer: The value of 1,150 is 1 standard deviation to the left of 1,200. The value of 1,300 is 2 standard deviations to the right of 1,200. Using the Empirical Guideline 34% of the area is between 1,150 and 1,200. Using the Empirical Guideline 47.5% of the area is between 1,200 and 1,300. Therefore, there is $0.34 + 0.475 = 0.815$ or 81.5% of the area between 1,150 and 1,300. Therefore, $P(1,150 \le x \le 1,300) = 0.815$.

$$\Sigma^{\circ}$$

chapter eight

VARIATION IN REPEATED SAMPLES-SAMPLING DISTRIBUTIONS

I know of scarcely anything
so apt to impress the imagination as the wonderful form
of cosmic order expressed by the
"Law of Frequency of Error" (the normal distribution).

– Sir Francis Galton

INTRODUCTION

In chapter 7 we were introduced to a continuous random variable. We then looked at the standard normal distribution and used the standardized variable, z, to find the area under the **standard normal curve**. We were also able to convert a normal distribution involving a continuous random variable x into a standard normal distribution. We then used the value of the standardized variable, z, to find an area under the standard normal distribution curve.

In this chapter we extend the idea of the probability distribution of a continuous random variable x to the probability distributions of the sample statistic $\bar{x}$. It is important to note that a sample statistic is a random variable and we refer to the probability distribution of the sample statistic as the **sampling distribution**. Using what is called the

Central Limit Theorem, we will be able to use the standardized variable, z, to transform the sampling distribution of $\bar{x}$ into a standard normal distribution. The Central Limit Theorem will tell us how to find the **mean and variance of the sampling distribution.** We will then be able to find the probabilities for values of $\bar{x}$ in an appropriate interval.

As an example, suppose we were interested in the distribution of car weights and want to find the probability that the average weight of 100 randomly selected cars is greater than 2,000 pounds. We need to know a value for the population mean, μ. For the present we will be given the value of the population mean. But later on this value will be obtained from a hypothesis or belief that someone has about the distribution of car weights. Once we have the value of μ, we use the Central Limit Theorem to find the mean and standard deviation of the sampling distribution of $\bar{x}$. We then find the z score for $\bar{x}$ and find the desired probability by finding the area under the appropriate tail of the standard normal curve.

The Central Limit Theorem states that for large sample size, the sampling distribution of the sample mean is approximately normal irrespective of the shape of the population distribution, and

$$E(\bar{x}) = \mu_{\bar{x}} = \mu \text{ and } sd(\bar{x}) = \sigma_{\bar{x}} = \frac{\sigma}{\sqrt{n}}.$$

The beauty of the Central Limit Theorem is that no matter what the distribution of the continuous variable x, the sampling distribution of $\bar{x}$ will be approximately normal as long as $n > 30$.

The steps outlined in the above example become the essence of the hypothesis testing we do in the chapters that follow. It will be very important that you understand what the Central Limit Theorem states.

CHAPTER 8 HOT SPOTS

1. **A Statistic Versus A Parameter.**
   Starts on **page 8-4**. J&B 8.1

2. **The Difference Between** $P(x \geq 65)$ **And** $P(\bar{x} \geq 65)$.
   Starts on **page 8–5**. Problems on **pages 8–9, 8–10, 8–11.**
   J&B 8.3

If you find other HOT SPOTS, write them down and use them as a focus of your discussions in the study group. Or you can use the HOT SPOT as the topic for a help session with your professor.

Hot Spot #1 – A Statistic Versus A Parameter

A very easy way to distinguish between a statistic and parameter is to remember that the word sample starts with an s just like statistics starts with an s. In the same way, both population and parameter start with the letter p. In fact, it might be a good idea to get into the habit of saying the phrases "sample statistic" and "population parameter" rather than the words statistics and parameter. Admittedly, this oversimplifies the problem, but a little association may go a long way to keeping the terms in perspective. On a more serious level, we need to look at the relationship between a sample statistic and a population parameter in both descriptive and inferential statistics.

In the branch of descriptive statistics, it is important to understand that we seldom know the value of a parameter. We are trying to estimate the parameter which is a measurable characteristic of the population. Since most populations are quite large, it would be difficult, if not impractical, to measure a characteristic of the population. Therefore, we use the sample statistic as an estimate of the population parameter. In the beginning, we will call this estimate a point estimate. Later on, we will extend our concept of an estimate to what we will call an interval estimate.

We have a different situation in inferential statistics where we assume a value of the population parameter. This is part of the inferential area labeled hypothesis testing. The hypothesis is that the population parameter is equal to the assumed value. The sample statistic is calculated and compared to the hypothesized value of the population parameter. If the sample statistic differs greatly from the assumed value of the population parameter, we say that we reject the hypothesis. Much of our work in the following chapters is an effort to quantify the process of hypothesis testing. As we do so, it becomes very important to distinguish between the value of the sample statistic and the value of the population parameter.

Hot Spot #2 – The Difference Between $P(x \geq 65)$ **And** $P(\bar{x} \geq 65)$

✳
Johnson And
Bhattacharyya
Section 8.3

One area of difficulty in this chapter is working with the Central Limit Theorem. It is fairly easy to know that the following two statements are part of the CLT:

$$E(\bar{x}) = \mu_{\bar{x}} = \mu \text{ and } sd(\bar{x}) = \sigma_{\bar{x}} = \frac{\sigma}{\sqrt{n}}$$

However, the real problem is to understand the effect these two statements have on a problem we are doing. As an example, we will look at the two questions from our example using the continuous random variable x = the speed of a car where:

$$\mu = 55 \text{ m.p.h. and } \sigma = 5 \text{ m.p.h.}$$

The first question involves finding the probability that a single speed is greater than 65. We write the first question as

$$P(x \geq 65)$$

The second question involves finding the probability that the average speed from a sample of n cars is more than 65. We write the second question as

$$P(\bar{x} \geq 65)$$

In order to understand the difference in these two questions, we need to first look at the distribution of speeds. Figure 8–1 shows the bell-shaped curve. The area shaded represents the 95% of the speeds between 45 m.p.h. and 65 m.p.h. (using the Empirical Rule).

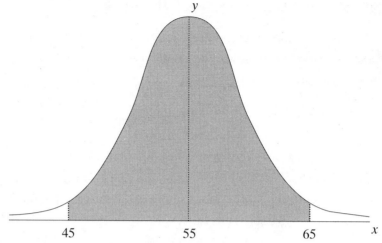

Figure 8–1:
95% Of Speeds
Between
45 M.P.H. and
65 M.P.H.

If we take a sample of speeds from this distribution, it will be unusual to get speeds greater than 65 m.p.h.. In fact, we will only get speeds greater than 65 m.p.h. approximately 2.5% of the time. We can now answer the first question as follows:

$$P(x \geq 65) = 0.025$$

Therefore, 97.5% of the time we would get speeds less than 65 m.p.h.. As a result, the average speed of any sample we take should usually be less than 65 m.p.h.. The difficult part of the question is to quantify the "usually less". This is where we make use of the Central Limit Theorem. The CLT tells us that the standard deviation of the distribution of the $\bar{x}$'s that we get from sampling has a value of 5 m.p.h. divided by the square root of the sample size. If we take a sample of 25 speeds, the standard deviation of the $\bar{x}$'s (sample means) will be $\frac{5}{\sqrt{25}}$.

$$\sigma_{\bar{x}} = 1 \text{ m.p.h..}$$

The bell-shaped curve for the distribution of $\bar{x}$ will be centered at $\mu = 55$, but the standard deviation will now be 1. The curve is shown below: (only the right half of the curve is shown to save space)

Figure 8–2:
Bell Curve
For Sampling
Distribution
Of Means

As you can see, there is very little of the area under the curve to the right of 65 m.p.h.. We are not able to use the Empirical Rule to find our answer to the second question $P(\bar{x} \geq 65)$, but we can find a z value and use either the z table or the HP-21S calculator to find our answer. The z value is found as follows:

$$z = \frac{65 - 55}{1} = 10$$

Unfortunately our z table only goes up to 5. The probability that z is greater than 5 is given below:

$$P(z \geq 5) = 0.5 - 0.4999997 = 0.0000003$$

Therefore the probability that z is greater than 10 must be smaller than the number we just found. If we use the HP-21S with the key that gives us the area under the normal curve, we find the exact answer of $P(z \geq 10) = 7.62 \times 10^{-24} = 0.0^{+}$.

We have found that although we might expect to occasionally find one speed selected from the population to be greater than 65, it would be extremely unusual to find a mean of 25 speeds selected from the population to be more than 65. Indeed, this result is the heart of what we do next in the chapter on hypothesis testing. Whenever we get a value that is rare, we consider it to be significant. If we do get such a rare value, we conclude that our original assumption (hypothesis) that μ was 55 m.p.h. is in error and our hypothesis should be rejected.

The two problems we just finished, finding

$$P(x \geq 65) \text{ and } P(\bar{x} \geq 65)$$

make it look like we are working with separate problems, but there is actually a very important common structure to the two problems. Both problems use the z score. We can write the z score in a form that will be very useful as we continue working in the next chapters. We can treat the z score using a template found in spread sheets like LOTUS 1-2-3. We simply fill in the box with the appropriate symbol.

The same z formula is used for several different z score calculations, with one substitution for each z score.

$$z = \frac{[] - \mu_{[]}}{\sigma_{[]}} \text{ where the [] can be } \begin{array}{l} 1.\ x \\ 2.\ \bar{x} \end{array}$$

We are actually using the same formula each time, with the box representing the statistic we are using. Later on we will increase the number of choices for the box to include a difference of sample means, a sample proportion, a difference of sample proportions, and a correlation coefficient. The denominator of the z score is called the standard error of the statistic we are using. When the sample statistic is $\bar{x}$, $\sigma_{\bar{x}}$ is referred to as the standard error of the mean. We often refer to $\sigma_{\bar{x}}$ simply as the standard error.

As we know from the Central Limit Theorem, the original population does not have to be normally distributed to use the theorem for a distribution of sample means. Furthermore, we do not need to know the value of σ if the sample size is considered large, $n > 30$. If the sample size is large we can use the value of s, the sample standard deviation, in place of the value of σ. As long as $n > 30$ we can write the z score as follows:

$$z = \frac{\bar{x} - \mu}{\frac{s}{\sqrt{n}}}$$

At this point, the use of s for σ may not seem like an important issue; but if $n \leq 30$, we will have to adjust the formula because s will be an underestimate of σ. We will talk more about this in the chapter that covers small samples using what is called the t distribution.

Hot Spot #2 Sample Problem and Solution: Given a normal distribution of cholesterol readings with $\mu = 200$ and $\sigma = 20$ find the following:

A. $P(x \leq 180)$ B. $P(\bar{x} \leq 180)$ with $n = 100$

Solution:

(A.)   Step 1.   Graph the bell-shaped curve.

Figure 8-3: Cholesterol Readings

Step 2.   Find the z score for $x = 180$.

$$z = \frac{180 - 200}{20} = -1$$

Step 3. $P(x \leq 180) = P(z \leq -1) = 0.5 - P(-1 \leq z \leq 0)$
$$= 0.5 - 0.3413$$

$P(x \leq 180) = 0.1587$

(B.)   Step 1.   Graph the bell-shaped curve.

Figure 8–4:
Average
Cholesterol
Readings

Step 2.   Find the z score for $\bar{x} = 180$.

$$z = \frac{180 - 200}{\dfrac{20}{\sqrt{100}}} = -10$$

Step 3. $P(\bar{x} \leq 180) = P(z \leq -10) = 0.0^{+}$

Sample Problem and Answer: Given that x has a normal distribution with $\mu = 70$ and $\sigma = 10$ find $P(\bar{x} \geq 75)$ when $n = 100$.

Answer: $P(\bar{x} \geq 75) = P(z \geq 5) = 0.5 - 0.4999997 = 0.0000003$.

Sample Problem and Answer: Given that x has a normal distribution with $\mu = 110$ and $\sigma = 5$ find $P(x \leq 100)$.

Answer: $P(x \leq 100) = P(z \leq -2) = 0.0228$.

CHAPTER 8 DISCUSSION QUESTIONS

These questions may be used in your study group or simply as topics for individual reflection. Whichever you do, take time to explain verbally each topic to insure your own understanding. Since these questions are intended as topics for discussion, answers to these questions are not provided. If you find that you are not comfortable with either your answers or that your group has difficulty with the topic, take time to meet with your professor to get help.

1. What is a parameter?

2. What does the Central Limit Theorem say about the mean and standard deviation of the random variable $\bar{x}$?

3. What happens to $sd(\bar{x})$ as n gets larger?

4. What does $sd(\bar{x})$ equal?

5. What is the difference between a population parameter and the random variable $\bar{x}$?

6. What does $E(\bar{x})$ equal?

7. What is random sampling?

8. What is the shape of the sampling distributions of $\bar{x}$ if $n > 30$?

9. What is the z score for $\bar{x}$?

10. What is the difference between a population distribution and a sampling distribution?

CHAPTER 8 TEST

The following information is for problems 1-4: A sample of 100 values of x is randomly selected from a population where x has a normal distribution with $\mu = 85$ and $\sigma = 5.25$.

1.   Find $E(\bar{x})$.

2.   Find $sd(\bar{x})$.

3.   What is the value of the standardized variable z for $\bar{x} = 75$?

4.   Find the probability that $\bar{x}$ is less than 75.

5.   Given that x is a continuous random variable with $\mu = 20.5$,
      $\sigma = 1.25$, and a random sample of $n = 49$; find $P(19.5 \le \bar{x} \le 20.5)$.

Given the following distribution for x, all possible samples of size 2 are selected:

x	-1	0	1
$P(x)$	$\dfrac{1}{3}$	$\dfrac{1}{3}$	$\dfrac{1}{3}$

6.   Find μ.

7.   Find σ.

8.   Find the probability distribution of $\bar{x}$.

9.   Find $E(\bar{x})$.

10.   Find $sd(\bar{x})$?

CHAPTER 8 TEST Questions and Answers

The following information is for problems 1-4: A sample of 100 values of x is randomly selected from a population where x has a normal distribution with $\mu = 85$ and $\sigma = 5.25$.

1.   Find $E(\bar{x})$.

 Answer: $E(\bar{x}) = \mu = 85$.

2.   Find $sd(\bar{x})$.

Answer: $sd(\bar{x}) = \dfrac{\sigma}{\sqrt{n}} = \dfrac{5.25}{10} = 0.525$

3.   What is the value of the standardized variable z for $\bar{x} = 75$?

Answer: Use $z = \dfrac{\bar{x} - \mu}{\dfrac{\sigma}{\sqrt{n}}}$. $z = \dfrac{75 - 85}{\dfrac{5.25}{\sqrt{100}}} = -19.05$.

4.   Find the probability that $\bar{x}$ is less than 75.

Answer: $P(\bar{x} < 75) = P(z < -19.05) = 0.0000^{+}$

5.   Given that x is a continuous random variable with $\mu = 20.5$, $\sigma = 1.25$, and a sample of $n = 49$; find $P(19.5 \le \bar{x} \le 20.5)$.

Answer: Use $z = \dfrac{\bar{x} - \mu}{\dfrac{\sigma}{\sqrt{n}}}$. $\quad z = \dfrac{20 - 20.5}{\dfrac{1.25}{\sqrt{49}}} = \dfrac{-0.5}{0.17857} = -2.8$.

$$P(19.5 \le \bar{x} \le 20.5) = 2P(0 \le z \le 2.8) = 0.9948$$

Given the following distribution for x, all possible samples of size 2 are selected:

x	-1	0	1
$P(x)$	$\dfrac{1}{3}$	$\dfrac{1}{3}$	$\dfrac{1}{3}$

6. Find μ.

Answer: $\mu = 0$

7. Find σ.

Answer: $\sigma = \sqrt{\dfrac{2}{3}}$

8. Find the probability distribution of $\bar{x}$.

Answer:

$\bar{x}$	-1	-0.5	0	0.5	1
$P(\bar{x})$	$\dfrac{1}{9}$	$\dfrac{2}{9}$	$\dfrac{3}{9}$	$\dfrac{2}{9}$	$\dfrac{1}{9}$

9. Find $E(\bar{x})$.

Answer: $E(\bar{x}) = 0$

10. Find $sd(\bar{x})$?

Answer: $sd(\bar{x}) = \sqrt{\dfrac{1}{3}}$

Σ°

chapter nine

DRAWING INFERENCES FROM LARGE SAMPLES

It is a capital mistake to theorize
before one has data.

– Arthur Conan Doyle

INTRODUCTION

In the last chapter we worked with continuous probability
distributions. The z score (standardized variable) that was first
presented in chapter 7 as a measure of position took on greater
importance. We used the standardized variable, z, and the standard
normal curve to answer probability questions related to the distributions
of x and $\bar{x}$. In this chapter we will use the z score to extend our work to
confidence intervals, estimating sample size, and hypothesis tests.

The **confidence interval** is an extension of the work we did in
chapter 2, where we found what are called **point estimates** of the
population parameter. For example $\bar{x}$ was a point estimate of the
population mean. We will now use both the $\bar{x}$ and the z score to find an
interval that contains μ. More importantly, we will be able to state a
level of confidence that the interval we create contains μ a certain
percent of the time.

In our work with the Central Limit Theorem in the last chapter
we assumed that n would be greater than 30. In this chapter we will

start with the assumption that we want our estimate of the population parameter to have certain accuracy. We then find the size of the sample that will guarantee the degree of accuracy we want in our interval estimate of the population parameter.

The second major part of this chapter extends our work to what is described as **hypothesis testing**. The idea is simply an extension of what we did in the last chapter with a z score. For example, we used the z score to find the probability that the sample mean was greater than some value. In each problem we were given the value of μ and σ to use in our calculation of z.

In this chapter we are given μ as a hypothesized value someone believes to be true. As we did before, we find the probability that $\bar{x}$ is larger than some value or less than some value by first finding the z score and then using the z table or calculator to find the probability. Traditionally, if the probability we find is less than 5%, we decide to reject the **null hypothesis**. We will also state the percent of the time that we could be wrong. There are several new terms defined that relate to hypothesis testing. It is important to take the time to learn what the terms mean.

An important key to this chapter, and the chapters that follow, is the general structure of hypothesis testing. This process will be used in all of the remaining chapters and involves five parts:

1. Formulate the **null and alternate hypotheses**, H_0 and H_1.
2. State the **test statistic** and the form of the **rejection region**.
3. With a specified value of α, determine the rejection region.
4. Calculate the value of the test statistic.
5. Make a decision to either reject H_o or fail to reject H_o.

A Study Guide For Statistics

CHAPTER 9 HOT SPOTS

If you find other HOT SPOTS, write them down and use them as a focus of your discussions in the study group. Or you can use the HOT SPOTS as the topic for a help session with your professor.

*

Johnson And
Bhattacharyya
Sections 9.2, 9.3

Hot Spot #1 – The Difference Between A Point And An Interval Estimate

Earlier in the text we used the sample mean, $\bar{x}$, to estimate the population mean, μ. The sample mean is called a point estimate of the population parameter. We need to take a sample from the population to obtain the data we use to calculate the sample mean. If the size of the sample is large, we feel good about our estimate. On the other hand if the sample size is small, we may feel uneasy about our point estimate of the parameter. In either situation, we do not have any way of finding out how accurate we are in our estimate.

In contrast, the confidence interval gives us a way of estimating a population parameter that includes a measure of how confident we are and the maximum error of our estimate. Constructing a confidence interval involves the same ideas we used when we worked with the Empirical Guideline with a known value of μ to find an interval that contained $\bar{x}$ 95% of the time. We simply reverse the process in finding a confidence interval. We start with a known value of $\bar{x}$, and find an interval that will contain the true value of μ 95% of the time.

We will first look at using the Empirical Guideline with the results of the Central Limit Theorem to find an interval that contains the sample mean, $\bar{x}$, 95% of the time. In this case we know the value of μ. As an example let x be a continuous random variable that represents SAT scores for a normal population with

$$\mu = 1,200 \text{ and } \sigma = 100 \text{ and } n = 100.$$

The Empirical Guideline tells us that 95% of the sample means should be in the interval

$$\mu \pm \frac{2\sigma}{\sqrt{n}}.$$

The interval includes sample means from $1,200 - \dfrac{2(100)}{\sqrt{100}}$ to $1,200 + \dfrac{2(100)}{\sqrt{100}}$

The interval is $(1{,}180\,,\,1{,}220)$ and is illustrated in the following graph:

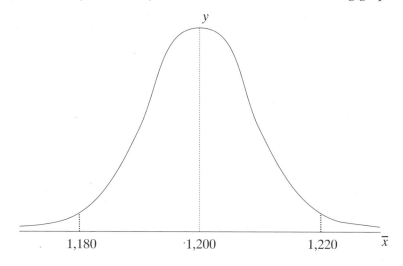

Figure 9–1:
Interval from
1,180 to 1,220

In contrast, when we construct a confidence interval, we do not know the value of μ. So we find $\bar{x}$ and use the idea of a z score to create an interval that contains μ. We start with the z score:

$$z = \frac{\bar{x} - \mu}{\dfrac{\sigma}{\sqrt{n}}}$$

We want what we will call the 95% confidence interval for estimating the population mean, μ. We will be 95% confident that the interval we construct from $\bar{x}$ will contain the true value of μ. In this sense we will accept being wrong 5% of the time. The z score for the 95% confidence interval is found by getting a value for $z\left(\frac{0.05}{2}\right)$. When $z = 1.96$, we have 2.5% of the area under the normal curve in the right tail. So we will use $z \pm 1.96$ to get our 95% confidence interval.

We now substitute for z and rewrite the z score. Solving for μ we get the following:

$$\mu = \bar{x} \pm 1.96\frac{\sigma}{\sqrt{n}}$$

$$\mu = \bar{x} \pm Z\,\frac{\sigma}{\sqrt{n}}$$

Generally speaking, we do not know σ. So, we use the sample standard deviation s as an estimate for σ. Since we are using the Central Limit Theorem to find $\sigma_{\bar{x}}$, it is important that the sample size be large, $n > 30$.

Returning to our example on SAT test scores, we will consider the case where we have no idea of the value of μ. As an illustration we will assume that all we have is a sample with

$$\bar{x} = 1{,}210 \text{ and } s = 100 \text{ and } n = 100.$$

We want to find the 95% confidence interval for μ. So we substitute the values of $\bar{x}$, s, and n into the formula as follows:

formula
- $\mu = \bar{x} \pm 1.96 \dfrac{\sigma}{\sqrt{n}}$

substituting
- $\mu = 1{,}210 \pm 1.96 \dfrac{100}{\sqrt{100}}$

simplifying
- $\mu = 1{,}210 \pm 19.6$

interval
- $1{,}190.4 \leq \mu \leq 1{,}229.6$

We then say that we are 95% confident that the interval from 1,190.4 to 1,229.6 contains the population mean μ.

We usually do not know if the interval actually contains the true value of μ. We can only say we are 95% confident that the interval contains μ. However, to see how the confidence interval works, we assumed we knew that $\mu = 1{,}200$. In the above example, the confidence interval we constructed did contain $\mu = 1{,}200$.

Now, let us see what it means to be 95% confident that the interval contains μ. We need to return to what we did with the Empirical Guideline, where we used a value of 2 rather than the more accurate z score of 1.96. We found that if $\mu = 1{,}200$ then 95% of the sample means would be in the interval from 1,180 to 1,220. If we use any

$\bar{x}$ in this interval to construct the confidence interval for μ, the confidence interval will contain μ.

The 95% confidence interval for μ we constructed from $\bar{x} = 1,210$ contained the value of 1,200. We need to note that $\bar{x}$ was in the interval from 1,180 to 1,220 that we obtained from the Empirical Guideline. In fact, if we construct the confidence interval using any $\bar{x}$ from the interval $(1,180 , 1,220)$, the confidence interval will contain the value 1,200.

Only 5% of the $\bar{x}$'s are outside of the interval from 1,180 to 1,220. If we use a $\bar{x}$ outside the interval $(1,180 , 1,220)$, the confidence interval we construct will not contain 1,200. To show this we will use $\bar{x} = 1,175$ and $s = 100$ and $n = 100$.

- $\mu = \bar{x} \pm 1.96 \dfrac{\sigma}{\sqrt{n}}$ *formula*

- $\mu = 1,175 \pm 1.96 \dfrac{100}{\sqrt{100}}$ *substituting*

- $\mu = 1,175 \pm 19.6$ *simplifying*

- $1,155.4 \le \mu \le 1,194.6$ *interval*

This time the 95% confidence interval goes from 1,155.4 to 1,194.6 and the confidence interval does **not** contain $\mu = 1,200$.

Figure 9-2: Illustration Of Choosing an $\bar{X}$ Outside The 95% Confidence Interval

The following table helps summarize the results of the examples:

Table 9–1:
Summary Of
Analysis Of
Confidence
Interval

Summary
1. $\bar{x}$ is in the interval $(1,180\,,1,220)$ 95% of the time.
2. We use an $\bar{x}$ in the interval to construct the confidence interval.
3. The confidence interval contains $\mu=1200$.
1. $\bar{x}$ is **not** in the interval $(1,180\,,1,220)$ 5% of the time.
2. We use an $\bar{x}$ outside the interval to construct the confidence interval.
3. The confidence interval does **not** contain $\mu=1200$.

Another way of saying what is in the table is that if:

1. We take 100 different random samples, with $n > 30$;
2. Find $\bar{x}$ and s for each sample;
3. Construct the confidence interval for each $\bar{x}$, then 95 of the 100 confidence intervals will contain $\mu = 1,200$.

Hot Spot #1 Sample Problem and Solution: Given the distribution of SAT scores discussed above with $\bar{x} = 1,210$, $s = 100$, and $n = 100$ show that the 95% confidence interval contains $\mu = 1,200$.

Solution: Step 1: Substitute values of $\bar{x}$, s, and n into $\mu = \bar{x} \pm 1.96\dfrac{\sigma}{\sqrt{n}}$.

$$\mu = 1,210 \pm 1.96\frac{100}{\sqrt{100}}$$

Step 2. Determine the upper and lower values of the interval
$\mu = 1,210 \pm 19.6$.

$1,190.4 \le \mu \le 1,229.6$

Step 3. The interval from 1,190.4 to 1,229.6 contains $\mu = 1,200$.

Sample Problem and Answer: Given the distribution of SAT scores discussed above with $\bar{x} = 1,250$, $s = 100$, and $n = 100$ show the 95% confidence interval does **not** contain $\mu = 1,200$.

Answer:

- $\mu = \bar{x} \pm 1.96 \dfrac{\sigma}{\sqrt{n}}$ *formula*

- $\mu = 1,250 \pm 1.96 \dfrac{100}{\sqrt{100}}$ *substitute*

- $\mu = 1,250 \pm 19.6$ *simplify*

- $1,230.4 \le \mu \le 1,269.6$ *interval*

The confidence interval does **not** contain $\mu = 1,200$.

Sample Problem and Answer: Given the following five values of $\bar{x}$ for the SAT scores discussed above, a 95% confidence interval is to be constructed for each $\bar{x}$. Which one of the values of $\bar{x}$ will result in the confidence interval containing $\mu = 1,200$?

     1,195   1,235   1,218   1,100   1,215

Answer: All the values, except 1,100 and 1,235, will result in a 95% confidence interval that contains $\mu = 1,200$.

*Johnson And
Bhattacharyya
Section 9.4*

Hot Spot #2 – Deciding On The Null Hypothesis, H_0.

It is fairly easy to decide on the null hypothesis when doing our own research. All we do is let the alternate hypothesis equal what we want to prove. We often refer to the alternate hypothesis as the research hypothesis. The null hypothesis then becomes the complement of the alternate hypothesis or research hypothesis. For example, if we want to prove that the population mean of highway speeds is greater than 55, we write

$$H_a: \quad \mu > 55.$$

The null hypothesis then becomes the complement of > 55.

$$H_0: \quad \mu \leq 55.$$

We usually write the null hypothesis using the = sign as follows: (**Hot Spot #5** discusses why we use the equal sign).

$$H_0: \quad \mu = 55.$$

Technically, a null hypothesis can be written with any one of the following three symbols:

$$\leq \quad \text{or} \quad = \quad \text{or} \quad \geq$$

Remember: The null hypothesis has the equal sign in it.

Rather than making it difficult to determine which form to use in the null hypothesis, as a rule of thumb, we just remember that **the null hypothesis has the equal sign in it**.

The alternate hypothesis (called the research hypothesis) then has one of the following symbols in it:

$$< \quad \text{or} \quad \neq \quad \text{or} \quad >$$

(**Hot Spot #4** discusses how to decide on the choice of symbols for the alternate hypothesis). The choice of the alternate hypothesis really is just a matter of deciding whether the critical region is in one or two tails of the standard normal curve.

Hot Spot #3 – Deciding On The Test Statistic

Johnson And
Bhattacharyya
Sections 9.4, 9.5

In this chapter, we use the test statistic z in our test of hypothesis. The sample statistic we use with the z test can be in one of three forms. We can use the sample mean ($\bar{x}$), the sample proportion ($\hat{p}$), or the number of successes (x). The choice of the test statistic is made as we read the problem. We are looking for two words, mean and proportion. When a problem involves the proportion we often have to interpret the meaning from the way the problem is written. One of the keys is to look for a percent that is given; the value given will usually be the value of the population proportion used in H_0.

When we read that we are trying to show something is significant about the population mean, we use

$$z = \frac{\bar{x} - \mu}{\frac{\sigma}{\sqrt{n}}}$$

If we read that we are trying to show something significant about the population proportion, we use

$$z = \frac{x - np}{\sqrt{npq}} \quad \text{or} \quad z = \frac{\hat{p} - p}{\sqrt{\frac{pq}{n}}}$$

The second of these two formulas for z has the advantage of fitting the structure we used earlier:

$$z = \frac{[] - \mu_{[]}}{\sigma_{[]}} \quad \text{where the } [] \text{ is } \begin{array}{l} \text{1. } x \\ \text{2. } \bar{x} \\ \text{3. } \hat{p} \end{array}$$

Expanded
Template For z
Test Statistic

We use the z test statistic to make inferences about the population parameter. As we go into the following chapters we will continue to use the z value to test hypotheses about the population parameter.

❋

Johnson And
Bhattacharyya
Section 9.4

Hot Spot #4 – Deciding On A One-Tailed Versus A Two-Tailed Test

The research hypothesis is what we want to prove. We write the research hypothesis as what we call the alternate hypothesis, H_a. We use one of the following three symbols of inequality in H_a:

$$< \text{ or } \neq \text{ or } >$$

We look for key words in the problems we do to decide on the choice of the inequality symbol to use with H_a. The following table shows many of the words that are used with each of the inequality symbols. In each situation the statement in the problem would involve showing that the population parameter was significantly _____ , where the blank would be one of the words taken from the list that follows:

Table 9–2:
Words And
Symbols Used In
The Alternate
Hypothesis

Table of Words and Symbols Used in H_a

$<$	$\neq$	$>$
Less Than	Different	More Than
Decreased	Changed	Increased
Reduced		Greater
Lowered		Higher
Worse		Better

One of the important points to remember is that the direction of the inequality symbol tells us the location of the critical region. The inequality symbol is like the directional signal in a car. When the arrow points left <, we have the critical region on the left side of the curve. When the arrow points right >, we have the critical region on the right side of the curve. When the symbol is not equal ≠, it is like having the emergency flasher system on where both tail lights blink. We then have the critical region in both tails of the curve.

Hot Spot #5 – Why We Use The = In A One-Tailed Test

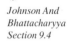
*Johnson And
Bhattacharyya
Section 9.4*

It is easiest to look at this question through an example using highway speeds. In a hypothesis test of $H_{0:}$ $\mu \leq 55$ and $H_{a:}$ $\mu > 55$ with $\sigma = 5$, a sample of $n = 25$ gave a sample mean of 57. We are interested in the probability that $\bar{x}$ is greater than 57 given that $\mu \geq 55$. We can start by looking at the normal curve with $\mu = 55$. Using the z formula for $\bar{x} = 57$, we get $z = 2$. The area to the right of 57 is $P(z > 2) = 0.0228$, and represents 2.28% of the area under the curve. The probability equal to the area under the tail of the curve is called the significance level of the hypothesis test. We use the symbol α for the size of the significance level.

*Figure 9–3:
Significance
Level Of The
Hypothesis Test*

Next, we will look at the area under the normal curve for a value of $\mu < 55$, that is $H_{0:}$ $\mu = 53$. Using the z formula for $\bar{x} = 57$, we get $z = 4$. Using the HP-21S, we find that the area to the right of 57 is $P(z > 4) = 0.0000317$ and represents 0.00317% of the area under the curve. This time we say that $\alpha = 0.0032\%$. The table in Table 9–1 below shows the value of α for different values of μ using the HP-21S.

Table 9–3:
Significance
Levels For
Different
Population
Mean Values

Values of α for Different Population Mean Values	
μ	α
55	2.28%
54	0.13%
53	0.0032%
52	0.000029%
51	0.00000010%

When we compare values in the table, we see that for values of $\mu < 55$, the area under the tail of the curve is much less than the area for $\mu = 55$. In this sense, the value of μ that is given by the equal sign in the null hypothesis, H_0, gives us the worst case scenario for the α value; that is, the largest value of α. Thus, it is common practice to simply write the null hypothesis with the equal sign. In the above example we would write

$$H_0 : \mu = 55 \text{ and } H_a : \mu > 55.$$

Hot Spot #6 – Choosing A Value Of α

It is important that we recall that the value of alpha (α) is the probability of rejecting a true null hypothesis. When we reject H_0, we report our findings to others in the form of a research paper or news release. If the null hypothesis is true and we make the mistake of rejecting H_0 and reporting our results to the world, we will be very embarrassed to say the least. Most of us are very worried about making a fool of ourselves in public. So we try to keep the value of α small. We usually choose a value of 1% or 5% for α. There are times that we may select a value of 10% when we are in the early stage of a research program. It is often difficult to find significant results, and we may be willing to accept a large value of α.

There are other considerations that may affect the value of α. For example, if we plan to publish a research report in a journal, it may be a common and accepted practice that all research be conducted using a value of 5%. In other situations, we may be doing a research project and a committee or advisor may decide on the level of α that should be used.

Another point is involved when we choose the value of α. We will illustrate the point using a right-tailed hypothesis test. The smaller the value of α we choose, the larger the critical value. For example, in a one-tailed test using the z test statistic, the critical value is 1.645 for $\alpha = 5\%$, but it is 2.326 when $\alpha = 1\%$. In order to reject H_0 at 1%, we need a larger value of z. Since

$$z = \frac{\bar{x} - \mu}{\sigma_{\bar{x}}}$$

we will need a larger difference between $\bar{x}$ and μ. In research it is usually very difficult to find large differences between the sample mean and its hypothesized value. Thus, it makes more sense to use the smaller critical value of 1.645 to make it easier to reject H_0.

It helps to keep in mind the inverse relationship between α and the critical value:

α gets smaller as the critical value gets larger.

In summary, if we must choose the value of α on our own, we use 5%.

Hot Spot # 7 – The Relationship Between α And The Reported P Value.

There is a common practice these days to actually report the significance level at the value given by the test statistic. The following example illustrates the practice:

In a hypothesis test of $H_0: \mu = 55$ and $H_a: \mu > 55$ with $\sigma = 5$, a sample of $n = 25$ gave a sample mean of 57. Test for a significant increase at $\alpha = 5\%$.

The value of the test statistic z is

$$z = \frac{57 - 55}{\frac{5}{\sqrt{25}}} = 2$$

We then find the probability that z is greater than 2:
$$P(z > 2) = 0.0228$$
We say that the reported P value of α is 2.28%.

We now have several choices that we can make:

1. Since the critical value is 1.645, we can report our research findings significant at the predetermined value of $\alpha = 5\%$.

2. We can simply report our research finding at the reported P value of 2.28%.

3. We can combine the two approaches by arguing that the P value of 2.28% is less than $\alpha = 5\%$; hence, we reject H_0.

There are two disadvantages to the first choice. First, we are not able to pass on the additional information that the reported P value provides, namely how significant our result is. Secondly, there are times when a reported P value would allow us to report research findings that would otherwise go unreported. For example, if the sample mean in the above hypothesis test was equal to 56.5, the value of the test statistic would have been equal to 1.5; we would fail to reject at $\alpha = 5\%$. However, if we decide to use the reported P value, we find that the probability that z is greater than 1.5 is 0.0668. We could report a significant increase at 6.68% and leave it to the reader to determine if he or she wishes to accept the risk of the type I error.

In the second choice, we seem to have the advantage of not having to worry about a predetermined significance level. However, there is another consideration that we need to address. The value of α (the probability of rejecting a true null hypothesis) is related to the value of β (the probability of failing to reject a false null hypothesis). The two values are inversely related.

α goes down in value as β goes up in value.

Generally speaking, we do not worry too much about the value of β when we fail to reject the null hypothesis.

The type I error associated with α is considered the more serious error. However, β is related to the power of the test. The power of the hypothesis test is the probability of rejecting a false null hypothesis.

$$\text{Power} = 1 - \beta$$

As β gets larger, the power gets smaller. It should be clear that very small values of α ultimately affect the power. As α gets smaller, β gets larger, and power gets smaller.

It is not appropriate to provide quantitative comparisons of the three values at this point in the course. But it is important to understand that one should not allow α to be reported at very low values without considering the possible effect on the power of the test.

In the third choice we are able to combine the best of what is involved in the first and second choices. When we use the third choice, it is common practice to first state the significance level of the hypothesis test at $\alpha = 5\%$. We then find the reported P value and compare it to α. If $P < \alpha$ then we reject H_0. The reported P value then has the advantage of showing how significant the result is.

One of the more subtle advantages of using the P value with a stated level of significance is that we do not need to find critical values. If we are using $\alpha = 5\%$, then any value of P less than 5% will allow us to reject H_0.

❋

Johnson And Bhattacharyya Section 9.4

Hot Spot #8 – The Difference Between Reject And Fail To Reject

It is fairly easy to figure out what is means to reject the null hypothesis. Failing to reject is just the flip side of the coin. The statement "Fail To Reject" is **not**, however, the same as accepting the null hypothesis. When we reject H_0, we do so at the significance level called alpha, α. The significance level states the percentage of the time we are making a type I error by rejecting H_0, when it is in fact true.

When we make the statement that we "Fail To Reject H_0, we are not obligated to give an error analysis. We have not in effect made any decision about the null hypothesis. If we, however, decided to accept H_0, we would need to indicate the percentage of the time we are making an error. The error in accepting a false H_0 is called a type II error, and the probability of making the error is called beta, β. Even though we agree on a significance level (α) at the beginning of a hypothesis test, we seldom know the value of β. Therefore, when the test statistic does not fall in the critical region, the best we can do is say that we fail to reject H_0.

Hot Spot #9 – The Relationship Between α, β, And n

✳
*Johnson And
Bhattacharyya
Section 9.4, 9.5*

The relationship between α and β is essentially that as one goes up in value, the other goes down in value. We often say that α and β are inversely related. However, the relationship is **not** like inverse variation; the product does **not** equal a constant. Furthermore, the sum of α and β is not 1. It is better to think of the two values as the opposite ends of a "teeter-totter".

*Figure 9–4:
Illustration Of
The Relationship
Between alpha
and beta*

As we push down on one side of the "teeter-totter", that is make α larger (heavier), the β value will become lighter (smaller). It is this inverse relationship that causes problems when we make the value of α extremely small. The value of β will then become quite large. This would be a bad situation because the power of the test of hypothesis is equal to $1 - \beta$.

One of our real concerns is then how to lower both α and β. The solution to our problem is quite easy in one sense of the word. All we have to do is increase the sample size we use in getting the sample mean or proportion. As easy as it sounds, using a larger sample size creates a number of related problems. Large sample sizes require more time, money, and energy. And at times it simply is neither practical nor possible to use a large sample.

The following problem using the binomial distribution shows how the values of α, β, and n are related. We will use $n = 10$ and $p = 0.5$ and set up the hypotheses as follows:

$$H_0: p = 0.5 \text{ and } H_a: p < 0.5$$

We will use a decision rule to reject H_0 if $x \leq 3$. The probability of the type I error is determined by finding $P(x \leq 3)$ when $p = 0.5$ and $n = 10$.

$$\alpha = P(x \leq 3) = 0.1719$$

We need a specific value of $p < 0.5$ to find the value of β. We will choose $p = 0.2$ to illustrate the process. We need to find $P(x > 3)$ when $p = 0.2$ and $n = 10$ to find the probability of the type II error.

$$\beta = P(x > 3) = 1 - P(x \leq 3) = 1 - 0.8791 = 0.1209$$

Although β is at an acceptable level (Power $= 1 - \beta = 88\%$), α is too large. We let $n = 5 \times 10 = 50$ and use a decision rule to reject H_0 if $x \leq 15$; the ratio of 15 to 50 is the same as the ratio of 3 to 10. Now the probability of a type I error is $P(x \leq 15)$ when $p = 0.5$ and $n = 50$.

$$\alpha = P(x \leq 15) = 0.0033$$

The value of β is found by taking $P(x > 15)$ when $p = 0.2$ and $n = 50$.

$$\beta = P(x > 15) = 1 - P(x \leq 15) = 1 - 0.9692 = 0.0308$$

We now find that we have lowered both the α and β values as a result of increasing the value of n from 10 to 50. Increasing n to even higher values will lower the values of α and β even more.

Now let us see what happens if we leave n at 10 but lower the α value. We will use the decision rule to reject H_0 if $x \leq 2$. We get the probability of the type I error by finding $P(x \leq 2)$ when $p = 0.5$ and $n = 10$.

$$\alpha = P(x \leq 2) = 0.0547$$

We find the probability of the type II error by finding the $P(x > 2)$ when $p = 0.2$ and $n = 10$.

$$\beta = P(x > 2) = 1 - P(x \leq 2) = 1 - 0.6778 = 0.3222$$

Although we lowered the value of α from 0.1719 to 0.0547, we increase the value of β from 0.1209 to 0.3222. In contrast, when we used a larger value of n we were able to lower α from 0.1719 to 0.0033 and lower β from 0.1209 to 0.0308.

The error analysis we just went through was reasonably easy to do for two reasons. The use of the HP-21S made it easy to answer the probability questions and we used a specific value of $p < 0.5$ from the alternate hypothesis. Generally speaking, we do not know a specific value for the alternate hypothesis. As a result, we seldom go through an error analysis. We simply assume that if α is kept at 5% and $n > 30$, the value of β will be small enough for us to have a powerful test.

CHAPTER 9 DISCUSSION QUESTIONS

These questions may be used in your study group or simply as topics for individual reflection. Whichever you do, take time to explain verbally each topic to insure your own understanding. Since these questions are intended as topics for discussion, answers to these questions are not provided. If you find that you are not comfortable with either your answers or that your group has difficulty with the topic, take time to meet with your professor to get help.

1. What is the difference between a point estimate and an interval estimate of a population parameter?

2. How do you narrow the width of the confidence interval?

3. Why do we use $\hat{p} = \hat{q} = \frac{1}{2}$ when finding the sample size to estimate the population proportion p?

4. What is the difference between the null and the alternate hypothesis, and how do you choose the null hypothesis?

5. When do you reject the null hypothesis?

6. What is the difference between a type I and type II error?

7. Why is it important to graph the curve and locate the critical value(s) on the standard normal curve when doing a test of hypothesis?

8. What is the difference between a one-tailed and two-tailed test of hypothesis?

9. If x is a continuous random variable that may not have a normal distribution, why is it important that $n > 30$?

10. What is the difference between α and the reported P value?

A Study Guide For Statistics

CHAPTER 9 TEST

1. If you want a smaller confidence interval for μ, what two choices do you have?

2. A researcher wants to prove that the average cholesterol levels of athletes is less than 180. State the null and alternative hypothesis.

3. Given $H_0: \mu = 120$ and $H_a: \mu < 120$, what is the critical value for a hypothesis test at $\alpha = 0.05$?

4. Given the accepted view that test scores average 1,200, a sample of 50 scores gave $\bar{x} = 1,237$ and $s = 125$. Test for a significant difference at $\alpha = 0.05$.

5. Find the 95% confidence interval for the population mean in #4.

6. What is the value of E, the maximum error, of the estimate in #4.

7. A hypothesis test of $H_0: \mu = 55$ and $H_a: \mu > 55$ is done with $\alpha = 0.05$. The value of the test statistic is $z = 1.79$. Use the reported P value to complete the hypothesis test.

8. What sample size is necessary to estimate the population proportion within 0.01 with 95% confidence?

9. A random survey of 100 managers found that 35 of them favored a national advertising campaign. Is this a significant change from the accepted view that 40% of the managers support the campaign? Test at $\alpha = 0.05$..

10. Find a 95% confidence interval for the population proportion, p, in #9.

CHAPTER 9 TEST Questions and Answers

1. If you want a smaller confidence interval for μ, what two choices do you have?

Answer: Either increase the sample size or increase the value of α.

2. A researcher wants to prove that the average cholesterol levels of athletes is less than 180. State the null and alternative hypothesis.

Answer: $H_0: \mu = 180$ and $H_a: \mu < 180$.

3. Given $H_0: \mu = 120$ and $H_a: \mu < 120$, what is the critical value for a hypothesis test at $\alpha = 0.05$?

Answer: The critical value is -1.645.

4. Given the accepted view that test scores average 1,200, a sample of 50 scores gave $\bar{x} = 1237$ and $s = 125$. Test for a significant difference at $\alpha = 0.05$.

Answer: Step 1. $H_0: \mu = 1,200$ and $H_a: \mu \neq 1,200$.

Step 2. $\alpha = 0.05$.

Step 3. $z = \dfrac{1,237 - 1,200}{\dfrac{125}{\sqrt{50}}} = 2.09$.

Step 4. The critical value for a two-tailed test at $\alpha = 0.05$ is ± 1.96.

Step 5. We Reject H_0.

5. Find the 95% confidence interval for the population mean in #4.

Answer: The 95% confidence interval for μ is given by

$$1,237 - 1.96\left(\frac{125}{\sqrt{50}}\right) \le \mu \le 1,237 + 1.96\left(\frac{125}{\sqrt{50}}\right).$$

$$1,202.35 \le \mu \le 1,271.65.$$

6. What is the value of E, the maximum error, of the estimate in #4.

Answer: $E = 1.96\left(\frac{125}{\sqrt{50}}\right) = 34.65.$

7. A hypothesis test of $H_0 : \mu = 55$ and $H_a : \mu > 55$ is done with $\alpha = 0.05$. The value of the test statistic is $z = 1.79$. Use the reported P value to complete the hypothesis test.

Answer: $P = P(z > 1.79) = 3.67\%$. Since $P < \alpha = 0.05$, we Reject H_0.

8. What sample size is necessary to estimate the population proportion within 0.01 with 95% confidence?

Answer: $n = \left[\frac{1.96}{0.01}\right]^2 \cdot \frac{1}{4} = 9,604.$

Drawing Inferences From Large Samples

9.   A random survey of 100 managers found that 35 of them favored
      a national advertising campaign. Is this a significant change from
      the accepted view that 40% of the managers support the
      campaign? Test at $\alpha = 0.05..$

Answer: Step 1. $H_0: P = 0.40$ and $H_a: P \neq 0.40$.

         Step 2. $\alpha = 0.05$.

         Step 3. $z = \dfrac{\hat{p} - p}{\sqrt{\dfrac{pq}{n}}} = \dfrac{0.35 - 0.40}{\sqrt{\dfrac{(0.4)(0.6)}{100}}} = 1.02$.

         Step 4.   The critical value for a two-tailed test at $\alpha = 0.05$ is
                   ± 1.96.

         Step 5.   We Fail to Reject H_0.

10.   Find a 95% confidence interval for the population proportion, p,
      in #9.

Answer: $\hat{p} = 0.35$

$$\hat{p} - 1.96\sqrt{\frac{\hat{p}\hat{q}}{n}} \leq P \leq \hat{p} + 1.96\sqrt{\frac{\hat{p}\hat{q}}{n}}$$

$$0.35 - 1.96\sqrt{\frac{(0.35)(0.65)}{100}} \leq P \leq 0.35 + 1.96\sqrt{\frac{(0.35)(0.65)}{100}}$$

$$0.2565 \leq P \leq 0.4435.$$

$$\sum \circ$$

chapter ten

SMALL-SAMPLE INFERENCES FOR NORMAL POPULATIONS

When you can measure what you are speaking about
and express it in numbers you know something about it;
but when you cannot measure it,
when you cannot express it in numbers,
your knowledge is of a meager and unsatisfactory kind.

— Lord Kelvin

INTRODUCTION

The **Student's** t and **chi-square** distributions in this chapter
extend the work we have done with hypothesis testing in chapter 9. The
t distribution is used for hypothesis tests of μ when we have small
sample sizes. The chi-square (χ^2) distribution is used for hypothesis
tests involving the population variance, σ^2. It will help to remember
that in each case we will use the same five steps of hypothesis testing.

The t distribution is just an extension of what we have already
done with the z distribution. The extension is a necessary one. When we
work with small samples, we tend to get data from the center of a
distribution. The extreme values of a distribution often are not included
in the sample because 95% of the time we get values within two

standard deviations of the mean. The result is that the value of the sample standard deviation may well be a low estimate of σ. When we calculate the z score, the low estimate of σ can result in an inflated value of z. In fact, there is an inverse relation between sample size and the potential for an inflated value of z; the smaller the sample size, the greater the potential for inflation. The result can be a type I error.

The t distribution provides us a way of compensating for what may be an inflated value of the standard score. We use a new set of critical values that are adjusted for the smaller sample sizes. For example, a one-tailed critical value for the standard normal curve at 5% is 1.645. The comparable value for the t distribution for a sample of size 10 is 1.833; the value for t is increased to 2.13 when the sample size is reduced to 5. The adjusted critical values make it more likely that our rejection of the null hypothesis is a result of an actual change in the value of the population parameter rather than an inflated value of the standard score.

The chi-square distribution allows us to test hypotheses about the variance. The test statistic for the chi-square distribution is

$$\chi^2 = \frac{(n-1)s^2}{\sigma^2}.$$

Although the test statistic looks quite different from the standard score, we use the same five steps of hypothesis testing from chapter 9 to test a hypothesis about the population variance. The chi-square critical values, like the t values, are adjusted for sample size. But this time there is a direct relationship between the sample size and the critical value. The larger the sample size, the larger the critical value.
Again, it is worth noting that as we test hypotheses using either the t distribution or the chi-square distribution, we simply repeat the same five steps we used in chapter 9.

A Study Guide For Statistics

CHAPTER 10 HOT SPOTS

1. **Degrees Of Freedom.**
   Starts on **page 10–4.** J&B 10.2

2. **Difference Between t Test Statistic And z Test Statistic.**
   Starts on **page 10–6.** J&B 10.2

3. **Reporting P Values For The t Distribution.**
   Starts on **page 10–10.** Problems on **page 10–12, 10–13.**
   J&B 10.2, 10.4

4. **Reporting P Values For The χ^2 Distribution.**
   Starts on **page 10–13.** Problems on **pages 10–14, 10–15.** J&B 10.6

If you find other HOT SPOTS, write them down and use them as a focus of your discussions in the study group. Or you can use the HOT SPOT as the topic of a help session with your professor.

*Johnson And
Bhattacharyya
Section 10.2*

Hot Spot #1 – Degrees Of Freedom

The *t* and chi-square distributions require that we find what are called degrees of freedom before we can find critical values for hypothesis testing. The letters *df* are used to represent the degrees of freedom. In both the *t* and chi-square distributions the degrees of freedom value is equal to one less than the sample size.

$$df = n - 1$$

The formula makes it simple to find the degrees of freedom, but it does not provide us any insight into what the phrase means. The idea behind degrees of freedom is illustrated by the following example:

- We want 3 numbers that have a sum of 10.

- We are free to choose any two numbers we want. For example we could choose 2 and 3.

- The third number must be a number that provides a sum of 10. In our example, the third number must be a 5. In this sense we do not have any freedom relative to the third number.

In our example we only had two degrees of freedom,

$$df = 3 - 1 = 2$$

If we had started with five numbers that had a sum of 10, we would have been free to choose any four numbers. However, the fifth number would not be a free number. Once we select four numbers, the fifth number is determined by the sum. The degrees of freedom would be $5 - 1$.

In the same way, if we started with *n* numbers that have a sum of 10, we are only free to choose $n - 1$ numbers. The last number is again determined by the sum; so, degrees of freedom equals $n - 1$.

The important point about the above examples is the effect that the sum 10 had on the numbers we selected. We used a sum of 10 to illustrate what happens, but any sum could have been used. For that matter we could have used the mean rather than a sum. The restriction that the numbers have a particular mean has the same effect of reducing our degrees of freedom to $n - 1$.

In fact we have a similar loss of freedom for any set of numbers when the numbers are used in a calculation that must have a given value. It does not matter what calculation is done. The effect is the same. The sum was used in the above example because it is an easy calculation to do. The mean was used as a second example of a calculation that reduced the degrees of freedom because it involves a sum. However, we would have the same loss in degrees of freedom if we used the restriction that the numbers we select have a particular standard deviation.

One final point needs to be made about degrees of freedom that will be used in later chapters. The point involves what happens when we select numbers if more than one calculation is involved as a restriction. For example, if the numbers we select must be used in two calculations, we lose two degrees of freedom and write $df = n - 2$. We can illustrate the idea using a familiar example from beginning algebra to show how we would lose two degrees of freedom. The problem is usually done with a variable and an equation that is then solved. In this case the problem is easy enough to solve that we will just give the solution by using observation.

- We want two numbers that have a sum of 10 and a product of 24.

- There are only two numbers that meet the requirement. The numbers are 6 and 4. We do not have any freedom. We started with two numbers, but lost two degrees of freedom because the numbers had to have a particular sum and product. Therefore, $df = 2 - 2 = 0$.

The above idea can be generalized in the following way. If the numbers we select must be used in k calculations that restrict the numbers we select, we lose k degrees of freedom and write $df = n - k$. For example if we had a large sample that we divided into k groups and required that the mean of each group be a specific value, we then would have $n - k$ degrees of freedom.

Hot Spot #2 – Difference Between t Test Statistic And z Test Statistic

✳
*Johnson And
Bhattacharyya
Section 10.2*

The formula we use for the t test statistic looks like the z test statistic we used when we substituted the sample standard deviation s for σ, the population standard deviation. This was not a problem as long as the sample size was larger than 30. When the sample size is less than or equal to 30, the sample standard deviation tends to underestimate σ. This can happen because small samples do not often include extreme values from the tails of the distribution. The effect on z, the standard score, is to make the value larger than it should be. When this happens in a hypothesis test, we increase our chance of rejecting a true null hypothesis and making a type I error.

The way we avoid making unnecessary type I errors when we do hypothesis tests with small samples is to adjust the critical value for the sample size. For example, the critical value of z for a test of hypothesis at 5% is 1.645 for a one-tailed hypothesis test on the right side of the curve. But if the sample size is reduced to 10, the critical value is increased to 1.833. As the sample sizes become smaller we continue increasing the size of the critical value. For example, the critical value becomes 2.132 when $n = 5$ and 2.920 when $n = 3$.

We get the new critical values using what is called the student's t distribution. The shape of the curve for the t distribution is bell-shaped looking like the now familiar standard normal curve used with the z

A Study Guide For Statistics

table. But the curve for the t distribution is flatter at the center and the tails of the curve seem to float above the horizontal axis when compared to the standard normal curve. As the sample sizes get smaller, the curve becomes even more flattened. We say that the shape of the curve depends on what is called degrees of freedom. The degrees of freedom equal one less than the sample size. We use the letters df to represent degrees of freedom and write the following expression:

$$df = n - 1$$

Hot Spot #1 discusses degrees of freedom.

The graph below shows a general comparison between the standard normal curve for z and student's t distribution curve used for t:

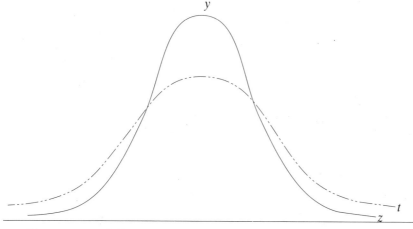

Figure 10–1: Comparison Of z Distribution To t Distribution

Actually, we get a different t curve for each value of df. As a result, it is more difficult to list the critical values for the t distribution. One way this could be done is to list a new table for each value of df. But this would require at least 30 tables just to do hypothesis tests. It is now common practice to simply list the critical values for the more common significance levels. The critical value could then be found by finding the appropriate degrees of freedom in a vertical column and then moving across to the required level of α. This allows us to get by with one table of values as opposed to carrying around a small book of tables.

There is a down side to using a single t table when it comes to reporting P values for a hypothesis test. This is discussed in **Hot Spot #3**, but an easy solution to the problem of using the t table is to use a calculator like the HP-21S. It makes it easy to both get a critical value for any degrees of freedom and report the P value for a hypothesis test. The following discussion assumes that we will be using the traditional t table to find critical values.

In the traditional t table, the critical values are listed for each of the commonly accepted significance levels for each value of df below 30. The table below shows the critical values of t for a 5% one-tailed hypothesis test as they would appear:

Table 10–1: Critical Values Of t For A 5% One-Tailed Hypothesis Test

Critical Values of t for a 5% One-Tailed Hypothesis Test

df	$t(0.05)$	df	$t(0.05)$
1	6.314	16	1.746
2	2.920	17	1.740
3	2.353	18	1.734
4	2.132	19	1.729
5	2.015	20	1.725
6	1.943	21	1.721
7	1.895	22	1.717
8	1.860	23	1.714
9	1.833	24	1.711
10	1.812	25	1.708
11	1.796	26	1.706
12	1.782	27	1.703
13	1.771	28	1.701
14	1.761	29	1.699
15	1.753		

A Study Guide For Statistics

It is common practice to simply use the value from the z table for values of df greater than 29. Therefore, we would use $t = 1.645$ for any $df > 29$. It should again be noted that calculators like the HP-21S now allow us to find the critical value for any degree of freedom and any value of α.

If we look carefully at the two curves shown below in Figure 10–2, we can see why the critical value is larger for the t test statistic. The two curves show the critical value for a 5% one-tailed test of hypothesis. The curve used for t is at 9 degrees of freedom. We can see that we must go out further in the tail of the t curve to get 5% of the area in the tail of the curve.

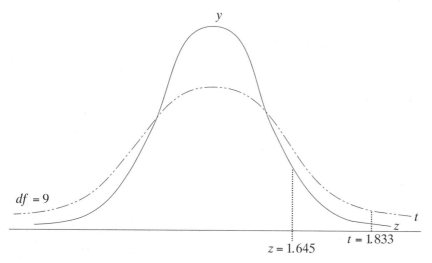

Figure 10–2:
5% Tails Of
z Distribution
Versus
t Distribution

The test statistic for the t distribution looks exactly like the z test statistic when we substitute s for σ.

$$t = \frac{\bar{x} - \mu}{\dfrac{s}{\sqrt{n}}}$$

We communicate that we are making the adjustment for a small sample by using the letter t for the standard score. Again, remember that we have to make the adjustment because:

1. We do not know the value of the population standard deviation.
2. The sample size is ≤ 30.

If we know σ, the sample size is no longer a problem. When we know σ, we can use the standard normal curve and find the critical value using the z table, regardless of the sample size.

✷
Johnson And Bhattacharyya Sections 10.2 and 10.4

Hot Spot #3 – Reporting P Values For The t Distribution

Reporting P values for a t distribution can be a problem when we do not use a calculator like the HP-21S. If we use the calculator, we can easily find the reported P value. All we have to do is enter the calculated value of t and the value for the degrees of freedom and press one key to give us the reported P value. However, P values can be a problem without the calculator.

The reason for the problem is that the table we use for critical values of t lists all the values of t for each degree of freedom under 30 for each value of α. The following table shows a typical row of values in a t table for $df = 9$:

Table 10–2: t Values For df=9

df	$t(0.005)$	$t(0.01)$	$t(0.025)$	$t(0.05)$	$t(0.10)$
9	3.250	2.821	2.262	1.833	1.383

The problem with reporting a P value for a calculated t test statistic is shown in the following example where a hypothesis test was used to find a significant reduction in cholesterol levels for a sample of 10 patients who reduced the amount of fat in their diets. The information used in the hypothesis test was as follows:

$$H_0: \mu = 200 \quad \text{and} \quad H_a: \mu < 200$$

$$\bar{x} = 192 \quad s = 11.35 \quad n = 10$$

$$t = \frac{192 - 200}{\frac{11.35}{\sqrt{10}}} = -2.229$$

Since the alternate hypothesis points to the left side of the curve, we need to think of the values in the above table as representing negative critical values on the left side of the curve. We then compare the calculated (observed) value of 2.229 to the entries in the table. Our observed value is between the critical values for α at $2\frac{1}{2}\%$ and 5%.

$$-2.262 < -2.229 < -1.833$$

We report the significance level as a value of less than 5% but more than $2\frac{1}{2}\%$.

$$2\frac{1}{2}\% < \alpha < 5\%$$

Using the HP-21S, we find that the actual P value is 0.0264.

Hot Spot #3 Sample Problem and Solution: A hypothesis test was done using the information that follows. Find the reported P value.

$$H_0: \mu = 1,200 \quad \text{and} \quad H_a: \mu > 1,200$$

$$\bar{x} = 1,214.75 \quad s = 23.14 \quad n = 20$$

$$t = \frac{1,214.75 - 1,200}{\frac{23.14}{\sqrt{20}}} \approx 2.851$$

Solution: Step 1. We calculate $df = 20 - 1 = 19$.

Step 2. We refer to the t table at 19 degrees of freedom and find the calculated value of t lies between the critical values of 2.539 at 1% and 2.861 at 0.5%.

Step 3. We report the significance level as less than 1% and greater than 0.5% and write $0.5\% < \alpha < 1\%$.

Sample Problem and Answer: A hypothesis test of $H_0: \mu = 55$ and $H_a: \mu \neq 55$ with $n = 10$ resulted in a value of $t = 1.842$. What is the reported P value?

Answer: Using $df = 9$, we find the calculated t value between 1.833 at 5% and 2.262 at $2\frac{1}{2}\%$. Since the hypothesis test was two-tailed we say that $\frac{\alpha}{2}$ is between $2\frac{1}{2}\%$ and 5%. So the hypothesis test was significant at a value of α between 5% and 10%.

Sample Problem and Answer: A hypothesis test of $H_0: \mu = 220$ and $H_a: \mu > 220$ with $n = 30$ resulted in a value of $t = 2.284$. What is the reported P value?

Answer: Using $df = 29$, we find the calculated t value between 2.045 at $2\frac{1}{2}\%$ and 2.462 at 1%, we say that the hypothesis test was significant at a value of α between 1% and $2\frac{1}{2}\%$.

Hot Spot #4 – Reporting P Values For The χ^2 Distribution

✳
Johnson And Bhattacharyya Section 10.6

The table used to get critical values for the chi-square distribution is similar to the t table. The critical values listed for each value of α are found by first finding the degrees of freedom. However, there is a very important difference in the tables when we find the critical value for the left side of the curve.

The chi-square distribution unlike the z and t distributions, is not symmetrical. Because of the non-symmetry we list the critical values for each of the two tails of the curve. It is no longer possible to simply use a ± sign in front of the value we found for the right side of the curve. When we look at the chi-square table we find a heading that gives areas to the right of the critical value for both sides of the curve. A typical heading and row of critical values for $df = 9$ follows:

df	$\chi^2(0.99)$	$\chi^2(0.975)$	$\chi^2(0.95)$	$\chi^2(0.90)$	$\chi^2(0.10)$	$\chi^2(0.05)$	$\chi^2(0.01)$
9	2.088	2.700	3.325	4.168	14.684	16.919	21.666

Table 10–3: Chi-Square Values For df=9

The values in the table should make it very clear that we cannot simply use the ± symbol to get the value for the left side of the curve. For example, at $\alpha = 0.05$ the critical value of χ^2 for the right side is 16.919, but the critical value for the left side is 3.325.

When we do a hypothesis test using the χ^2 table we find the critical value or values consistent with whether we have a one-tailed or two-tailed test. We, of course, also need the significance level. It is when we decide to do a hypothesis test using a reported P value that we run into the same difficulty we had with reporting P values with the t distribution.

Before we look at how we handle the problem with the chi-square table, it is again important to understand that we can avoid the problem using a calculator like the HP-21S. Once we have the degrees of freedom entered into the calculator, we can find the area to the right of our observed value by pressing one key. For that matter, we can get critical values for a hypothesis test just as easily. However, the remainder of the discussion assumes that we will use the traditional χ^2 table to report the P value associated with a hypothesis test.

As an example, we will do a hypothesis test on $H_0: \sigma^2 = 1.25$ and $H_a: \sigma^2 > 1.25$ at $\alpha = 0.05$ with $n = 10$. The sample gave $s^2 = 2.20$. When we calculate the test statistic we get the following:

$$\chi^2 = \frac{(n-1)s^2}{\sigma^2} = \frac{(9)(2.2)}{1.25} = 15.84$$

We then compare the calculated value to the table entries for $df = 9$. The calculated value is between 14.684 at $\alpha = 10\%$ and 16.919 at $\alpha = 5\%$. We would say that the hypothesis test was significant at a value between 5% and 10%. (Using the HP-21S, we find the exact value of $\alpha = 0.070$.)

Hot Spot #4 Sample Problem and Solution: Find the reported P value for the following hypothesis test on $H_0: \sigma^2 = 0.25$ and $H_a: \sigma^2 \neq 0.25$ with $s^2 = 0.5$ and $n = 10$.

Solution: $\chi^2 = \frac{(9)(0.5)}{0.25} = 18$ and $df = 10 - 1 = 9$.

18 is between 16.919 at $\frac{\alpha}{2} = 5\%$ and 19.023 at $\frac{\alpha}{2} = 2\frac{1}{2}\%$. The reported value of P is between 5% and 10%.

Sample Problem and Answer: Find the reported P value for a one-tailed hypothesis test on the left side of the curve if $\chi^2 = 5.22$ and $n = 15$.

Solution: We use $df = 15 - 1 = 14$.

5.22 is between 4.660 at $\alpha = 1\%$ and 5.629 at $\alpha = 2\frac{1}{2}\%$. The reported P value is between 1% and $2\frac{1}{2}\%$.

Sample Problem and Answer: Find the reported P value for a two-tailed hypothesis test if $\chi^2 = 50.475$ and $n = 30$.

Solution: We use $df = 30 - 1 = 29$.

50.475 is between 49.588 at $\frac{\alpha}{2} = 1\%$ and 52.336 at $\frac{\alpha}{2} = 0.5\%$. The reported P value is between 1% and 2%.

CHAPTER 10 DISCUSSION QUESTIONS

These questions may be used in your study group or simply as topics for individual reflection. Whichever you do, take time to explain verbally each topic to insure your own understanding. Since these questions are intended as topics for discussion, answers to these questions are not provided. If you find that you are not comfortable with either your answers or that your group has difficulty with the topic, take time to meet with your professor to get help.

1. When do we use a z test statistic for $n \le 30$?

2. What is the difference between a z table and a t table?

3. Explain the difference between the standard normal curve and the curve for the student's t distribution.

4. What is meant by degrees of freedom?

5. How does one report P values using (a) the t distribution (b) the chi-square (χ^2) distribution?

6. How does a t value compare to a z value for a given α?

7. How does sample size affect (a) the critical values of t and (b) the critical values of χ^2?

8. How does the curve for the chi-square (χ^2) distribution differ from the curve for the student's t distribution?

9. What assumptions must be met before we can use the student's t distribution?

10. What assumptions must be met before we can use the chi-square (χ^2) distribution?

CHAPTER 10 TEST

1. Find the critical values of t for a hypothesis test of $H_0: \mu = 55$ and $H_a: \mu \neq 55$ for $\alpha = 0.05$ with $n = 15$.

2. Find the critical value for a hypothesis test of $H_0: \sigma^2 = 1.25$ versus $H_a: \sigma^2 > 1.25$ for $\alpha = 0.05$ with $n = 20$.

Given the following information: $\bar{x} = 1,225$, $s = 35$ and $n = 25$

3. Test $H_0: \mu = 1,200$ and $H_a: \mu > 1,200$ for significance at $\alpha = 5\%$.

4. Give the reported P value.

5. Find the 95% confidence interval for μ.

Given the following information: $s^2 = \frac{1}{2}$ and $n = 20$

6. Test the hypothesis $H_0: \sigma^2 = \frac{1}{4}$ versus $H_a: \sigma^2 > \frac{1}{4}$ at $\alpha = 5\%$.

7. Give the reported P value.

8. Find the 95% confidence interval for σ^2.

$$\alpha = 0.05$$
$$\alpha = 0.025$$

Find the following probabilities:

9. $P(1.383 < t < 1.833)$

10. $P(32.852 < \chi^2 < 36.191)$

CHAPTER 10 TEST Questions and Answers

1. Find the critical values of t for a hypothesis test of $H_0: \mu = 55$ and $H_a: \mu \neq 55$ for $\alpha = 0.05$ with $n = 15$.

Answer: ± 2.145

2. Find the critical value for a hypothesis test of $H_0: \sigma^2 = 1.25$ versus $H_a: \sigma^2 > 1.25$ for $\alpha = 0.05$ with $n = 20$.

Answer: 30.144

Given the following information: $\bar{x} = 1,225$, $s = 35$ and $n = 25$

3. Test $H_0: \mu = 1,200$ and $H_a: \mu > 1,200$ for significance at $\alpha = 5\%$.

Answer: $t(0.05, 24) = 1.711$ $\qquad t = \dfrac{1,225 - 1,200}{\dfrac{35}{\sqrt{25}}} = 3.57$ $\qquad$ Reject H_0.

4. Give the reported P value.

Answer: Using the HP-21S, $P = 0.00077$.
$\qquad\qquad$ Using the t table, all we can say is that $\alpha < 0.005$.

5. Find the 95% confidence interval for μ.

Answer: $\qquad t(0.975) = -2.064 \qquad\qquad t(0.025) = +2.064$

$$1,225 - \frac{(2.064)(35)}{\sqrt{25}} < \mu < 1,225 + \frac{(2.064)(35)}{\sqrt{25}}$$

$$1,210.552 < \mu < 1,239.448$$

Given the following information: $s^2 = \frac{1}{2}$ and $n = 20$

6.   Test the hypothesis $H_0: \sigma^2 = \frac{1}{4}$ versus $H_a: \sigma^2 > \frac{1}{4}$ at $\alpha = 5\%$.

Answer: $\chi^2(0.05,19) = 30.144$ $\chi^2 = \dfrac{(19)\frac{1}{2}}{\frac{1}{4}} = 38$ Reject H_0.

7.   Give the reported P value.

Answer: Using the HP-21S, $P = 0.0059$.
          Using the χ^2 table all we can say is $\frac{1}{2}\% < \alpha < 1\%$.

8.   Find the 95% confidence interval for σ^2.

Answer: $\chi^2(0.975,19) = 8.907$ $\chi^2(0.025,19) = 32.852$
$$\frac{(19)\left(\frac{1}{4}\right)}{32.852} < \sigma^2 < \frac{(19)\left(\frac{1}{4}\right)}{8.907}$$
$$0.145 < \sigma^2 < 0.533$$

Find the following probabilities:

9. $P(1.383 < t < 1.833)$ where $df = 9$.

Answer: Both values of t are found on the line of the t table with $df = 9$.
          $P(1.383 < t < 1.833) = 0.10 - 0.05 = 0.05$.

10. $P(32.852 < \chi^2 < 36.191)$ where $df = 19$.

Answer: Both values of χ^2 are found on the line of the chi-square table
          with $df = 19$. $P(32.852 < \chi^2 < 36.191) = 0.025 - 0.01 = 0.015$.

$$\Sigma^{\circ}$$

chapter eleven

COMPARING TWO TREATMENTS

Knowledge is of two kinds.
We know a subject ourselves or
we know where we can find information upon it.

– Dr. Johnson

INTRODUCTION

In earlier chapters on hypothesis testing we used the value of a sample statistic to challenge an accepted value of a population parameter. We focused on both the mean and proportion. The goal was to show that there was a large enough difference between the value of the sample statistic and the population parameter that the difference could not be due to chance alone. When the difference was large, we argued that the value of the population parameter had changed. This approach, however, did not provide us any information about what was responsible for the change in the population parameter. In this chapter we introduce the idea of **experimental design** that refers to the manner in which subjects are chosen and assigned to treatment groups. We compare treatments using both independent and dependent samples to help us determine what is responsible for the change in the value of the population parameters.

Comparing Two Treatments

When we work with **independent samples**, the two groups are often referred to as control and experimental. In a more general sense we simply consider that we are comparing two treatments involving two populations. We then define a response variable that measures a characteristic that is affected by the treatment. The null hypothesis is that there is no treatment difference, i.e.. that the difference in the population parameters of the response variable is zero. We then compare the sample means; if the difference between the means is large enough, we argue that there is a significant difference in treatment.

Interestingly enough, variance becomes a focus in this chapter. When we compare two population means, we need to be concerned about the population variances. If one of the variances is much larger than the other, the value of our test statistic may then be inflated causing us to make a type I error.

The work with independent samples is an effort to determine the effect of a treatment on the population parameter. Unfortunately the difference in means may be a result of uncontrolled variables. One way of controlling the confounding variables (variables other than the response variable that may cause a difference in sample means) is to use **dependent samples** where data from the two samples is paired in a natural way. The test statistic used involves the differences of the paired data. The simplicity of the hypothesis test that the **mean difference** equals zero is complemented by the additional advantage of controlling the confounding variables.

In contrast, when we work with proportions, we are simply comparing two unknown population parameters. We seldom are worried about what may be responsible for the difference. We just want to know if there is a significant difference between population proportions. We state a null hypothesis that the two population values are equal. If the difference between the two sample proportions is large enough, we reject the null hypothesis in favor of the alternate hypothesis.

CHAPTER 11 HOT SPOTS

If you find other HOT SPOTS, write them down and use them as a focus of your discussions in the study group. Or you can use the HOT SPOT as the topic for a help session with your professor.

Johnson And Bhattacharyya Section 11.2

Hot Spot #1 – How Unequal Population Variances Can Cause Type I Errors

When we do a hypothesis test about the means of independent samples with n_1 or $n_2 \leq 30$, we must check to see if the variances of the two populations are equal. If we reject the hypothesis that population variances are equal, we then test for the null hypothesis that $\mu_1 = \mu_2$ using the degrees of freedom given by the formula:

$$df = \text{ the smaller of } n_1 - 1 \text{ or } n_2 - 1$$

This value is typically smaller than $n_1 + n_2 - 2$. The critical point is that the result is a more conservative hypothesis test of $\mu_1 = \mu_2$ because the critical value for the t test statistic gets larger as the degrees of freedom get smaller. The reason we need to be more conservative is illustrated in the following example.

We will consider two populations that have the same mean but very different variances, and hence different standard deviations. The graphs of the two populations are shown below.

Figure 11–1: Two t Curves With The Same Mean And Differing Standard Deviations

Since the standard deviation is larger for the top curve (Figure 11–1), it is possible that we could get much of our data from one of the tails of the distribution. The smaller standard deviation for the bottom curve (Figure 11–1) makes it likely that most of our data will come from the center of the distribution. As such, the difference in sample means could be relatively large resulting in an incorrect decision that we reject the null hypothesis of equal population means. This would result in a type I error. The summary table below (Table 11–1) shows how this might occur. The table shows that there is a small difference in sample means, equal sample size, and a rather large difference in the sample variance.

	n	$\bar{x}$	s^2
1	15	23.08	5.4275
2	15	21.90	1.7352

Table 11–1: Summary Of Sample Sizes, Means, and Variances For Two t Curves

The standard error of $\bar{x}_1 - \bar{x}_2$ is $\sqrt{\dfrac{5.4275}{15} + \dfrac{1.7352}{15}} = 0.691$.

When we calculate the value of the t test, we get $t = \dfrac{23.08 - 21.90}{0.691} = 1.708$.

Now we need to look at the one-tailed critical value we would use if we *assume that the variances were equal.* The critical value would be $t(0.05, 28) = 1.701$ and *we would reject the null hypothesis.* However, if we *assume unequal population variances* and use the more conservative critical value for $df = 14$, we get $t(0.05, 14) = 1.761$ and *we fail to reject.* If the populations variances were unequal, the more conservative value of t using df for the unequal population variance test would keeps us from making a type I error. If we did get a t value larger than 1.761, we could be confident that we were rejecting the null hypothesis because there was a difference in population means.

Two different things were at work in the analysis we just finished. The first involves the effect of the variances on the numerator of the test

Comparing Two Treatments

statistic. The second point relates to the effect of the variance on the denominator of the test statistic. Before we look in detail at these two points, it is important to emphasize that when we have unequal population variances we must be very cautious. This is true not only of our work with independent samples, but also of the work we will do later in what is known as analysis of variance (ANOVA). The following analysis assumes that the population means are actually equal.

Relative to the first point, we obtained a relatively low value of the sample mean from the first distribution because of the larger standard deviation. If the standard deviations had been equal, we would not have found so many values on the low end of one of the curves. The resulting sample means would have been closer in value and the numerator of the test statistic would be smaller. The effect would be a smaller value of the test statistic. Of course this does not take into account what happens in the denominator if the variances are equal.

The second point that we need to consider is the effect of the unequal population variances on the denominator of the test statistic. If the distributions both had the larger variance, the denominator would be larger and further reduce the size of the resulting t value. This of course, would be coupled with the reduction in the value of the numerator as explained above. The result of the two effects would be an even smaller value of the test statistic. In any case we would fail to reject the null hypothesis, unless there actually was a difference in population means.

The following calculation shows what happens to the t value if the numerator does not change and we use equal sample variances using the larger value.

The standard error of $\bar{x}_1 - \bar{x}_2$ is $\sqrt{\dfrac{5.4275}{15} + \dfrac{5.4275}{15}} = 0.851.$

When we calculate the value of the t test, we get $t = \dfrac{23.08 - 21.90}{0.851} = 1.387$.

  The result is a smaller value of t. We would fail to reject the null hypothesis in both of the hypothesis tests we did. The large variance for both populations makes it likely to get sample means that differ by a large amount and we do not find enough evidence to reject the null hypothesis of equal population means.

  If both distributions had the smaller variance, the denominator would become smaller. This would appear to increase the value of the test statistic. However, it would also be unlikely to get large differences in sample means when the variances are both small. Although it is more difficult to illustrate the effect in this case, if we did find a large value for the test statistic we would again be assured that it was a result of the treatment effect and not of a difference in population variance. The following calculation shows what happens to the t value if the numerator does not change and we use equal sample variances using the smaller value.

$$\text{The standard error of } \bar{x}_1 - \bar{x}_2 \text{ is } \sqrt{\dfrac{1.7352}{15} + \dfrac{1.7352}{15}} = 0.481.$$

When we calculate the value of the t test, we get $t = \dfrac{23.08 - 21.90}{0.481} = 2.453$.

  We would reject the null hypothesis in both of the hypothesis tests we did. The difference in means becomes more important with the smaller variances and results in a large t value. Our hypothesis was that the population means were equal making it highly unlikely that we would get a difference of means this large. So if we did get a difference this large, it would be very significant.

Comparing Two Treatments

✳
Johnson And
Bhattacharyya
Sections
11.4 and 11.5

Hot Spot #2 – How To Decide If The Sample Data Comes From Dependent Samples

Statistical analysis is much simpler when we collect our own data because we will know if we used dependent samples. It is when we are asked to do a hypothesis test for data that we did not collect that we have to make a decision. We must then decide if the data is from an independent sample or a dependent sample. There are two situations that indicate that we have independent samples. The first is when the sample sizes are not equal. We then know that we must have independent samples because dependent samples require the same sample size for both data sets. The second situation is when we are given a summary table that includes the means, the standard deviations, and the sample sizes. It again is very clear that we have independent samples because we do not calculate individual means and standard deviations for dependent samples.

It is when the data itself is given for the two groups that have equal sample sizes that we run into trouble. We must then search for a key word or phrase to help us out. Obviously a key word is "dependent". It makes it easy when the problem includes the information that we have been given two dependent samples. One of the key phrases to look for is "paired differences". Another phrase that may be a key is that the data represents "before and after" scores on some measuring instrument such as a test. However, there are times when there may be reasonable doubt even after looking for key phrases. We may need to look carefully at the description of how the experiment was conducted.

Dependent samples often involve the same subjects or objects. We then obtain paired data values that reflect the changes as a result of the treatment. The idea involved in a dependent sample is that by using the same subject we do not introduce confounding variables. Unfortunately there are times when it is not possible to use the same subject. For

A Study Guide For Statistics

example, in studies about heart disease we often need to match patients as best we can for the confounding variables such as weight, age, lifestyle, etc. We then measure the treatment effect on the variable of interest, for example cholesterol or blood pressure. The paired data for the matched subjects is then treated as a dependent sample. The pairing of subjects makes it easier to find a treatment difference if it exists.

Naturally, this brings us to the point of asking, "what happens if we still can not decide if the data comes from a dependent sample?". The answer to the question is to use the independent sample hypothesis test. At worst we will make a type II error. If we instead make the assumption that the data reflects an effort to control confounding variables using the dependent sample, we could make a type I error. The possible errors are discussed further in the next **Hot Spot** that examines the effect of dependent samples on the hypothesis test.

Hot Spot #3 – Why Is A Dependent Sample An Advantage In Hypothesis Testing?

Johnson And Bhattacharyya Section 11.5

The dependent sample gives us our first look at what is called blocking. We control confounding variables by matching subjects for the variables we want to control. The result is that we are more easily able to reject a false null hypothesis. The data set that follows   (Table 11–2) illustrates the point. The example involves twenty-four males aged 25–29 selected from the Framingham Heart Study. Each of the twelve pairs of data represents a smoker and a non-smoker who were matched with regard to age and physical characteristics. The random variable measured was systolic blood pressure.

Comparing Two Treatments

Smokers	Non-Smokers
122	114
146	134
120	114
114	116
124	138
126	110
118	112
128	116
130	132
134	126
116	108
130	116

Table 11–2: Blood Pressures Of Twenty-Four Males In The Framingham Heart Study

We are interested in testing the hypothesis that there is a difference in mean systolic blood pressure levels for the two populations. We will do the hypothesis test using the t test for dependent samples. We need to find the difference between each ordered pair. The differences are listed below:

Smokers	Non-Smokers	d
122	114	8
146	134	12
120	114	6
114	116	-2
124	138	-14
126	110	16
118	112	6
128	116	12
130	132	-2
134	126	8
116	108	8
130	116	14

Table 11–3: Ordered Pair Differences Of Blood Pressures In Framingham Heart Study

Using the differences as our random variable d, we find the following:

$$n = 12 \qquad \bar{d} = 6 \qquad s_d = 8.399134155$$

$$\frac{s_d}{\sqrt{12}} = 2.424621183$$

We are now ready to do the hypothesis test using the test statistic

$$t = \frac{\bar{d}}{\frac{s_d}{\sqrt{n}}}$$

Step 1. $H_0: \mu_S = \mu_{NS}$ versus $H_a: \mu_S \neq \mu_{NS}$.

Step 2.   The α level will be used at 5%.

Step 3.   The two-tailed critical value of t for $df = 12 - 1$ is ± 2.201.

Step 4. $t = \dfrac{6}{2.42462} = 2.47$

Step 5.   Reject H_0.

Now we will see what happens if we did not know that the data involved paired data. The following summary table gives the necessary values for the hypothesis test. We will assume that the variances for the two populations are equal.

Table 11–4: Necessary Values For Hypothesis Testing Of Twenty-Four Males In Framingham Heart Study

	n	$\bar{x}$	s^2
Smokers	12	125.67	78.4242
Non-Smokers	12	119.67	102.4242

Technically, with the variances equal and a small sample size, we should use the t test that uses the pooled standard deviation. Since the sample sizes are equal, we can use the following test statistic that will give us the same value:

$$t = \frac{\bar{x}_S - \bar{x}_{NS}}{\sqrt{\dfrac{s_S^2}{n_S} + \dfrac{s_{NS}^2}{n_{NS}}}}$$

Step 1. $H_0: \mu_S = \mu_{NS}$ versus $H_a: \mu_S \neq \mu_{NS}$.

Step 2. The α level will be used at 5%.

Step 3. The two-tailed critical value of t for $df = 24 - 2$ is ± 2.074.

Step 4. $t = \dfrac{125.67 - 119.67}{\sqrt{\dfrac{78.4242}{12} + \dfrac{102.4242}{12}}} = 1.55$

Step 5. We fail to reject H_0.

Perhaps we need to review what we have just completed:

We first did a hypothesis test *assuming the data represented a dependent sample*. We found the t value of 2.47 in the critical region. So *we rejected the null hypothesis, H_0.*

We then did a hypothesis test *assuming the data represented an independent sample*. We found the t value of 1.55 that was **not** in the critical region. Thus, *we failed to reject the null hypothesis.*

The important point in all this is that when we treated the data as a dependent sample, we found a significant difference. When we assumed the data came from an independent sample we did **not** find a significant difference. Clearly, it is to our advantage to use a dependent sample whenever we can.

A second point can also be made regarding what we should do if we do not know for sure that the data is from a dependent sample. We should then go ahead and treat the data as if it were from an independent sample. If we can reject the null hypothesis under the assumption of independence, we can be confident that we would come to the same conclusion if the sample were dependent. Simply stated, the independent sample hypothesis test is a more conservative position.

When we are unable to verify that the data comes from a dependent sample and thus use the test statistic for an independent sample, we only take the chance of failing to reject a false null (a type II error). However, if we incorrectly assume that the data is from a dependent sample and use the test statistic for dependent samples, we take the chance of rejecting a true null hypothesis (a type I error).

Hot Spot #4 – Deciding Which Test Statistic To Use For Independent Samples

✳
Johnson And
Bhattacharyya
Section 11.2

One of the more confusing things about this chapter is deciding on which test statistic to use for independent samples. The sense of confusion can be reduced to a very low level by looking at the decision process using a tree diagram like we used in probability.

Our first decision is to choose between a large and small sample case. If both of the sample sizes are more than 30, we can immediately find the value of the z test statistic and compare it to a critical value from the z table using the values of s_1^2 and s_2^2 for the values of σ_1^2 and σ_2^2. The first branch of the tree diagram can now be drawn. It is the upper branch of the tree diagram that follows in Figure 11–2 on the next page.

Comparing Two Treatments

If either sample is less than or equal to 30, we need to use a t test statistic and there are two choices based on whether or not the population variances are equal:

If the population variances are assumed to be equal, we use the pooled standard deviation form of the t test. It is drawn in Figure 11–2 as the top part of the lower branch.

If the population variances are not equal, we then use the test statistic with a more conservative degrees of freedom calculated from the formula shown in the diagram below. The formula is in Figure 11–2 as the bottom part of the lower branch.

The completed tree diagram is presented below in Figure 11–2. It is important to keep a visual image of the tree diagram as we look at a hypothesis test for an independent sample. The mental picture makes it easier to go through the decision process.

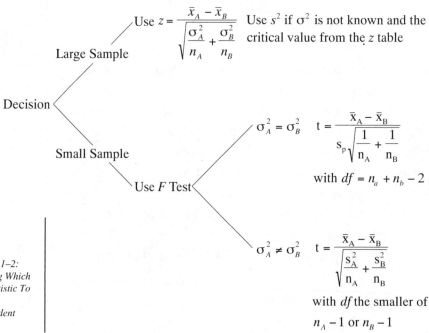

Use $z = \dfrac{\bar{x}_A - \bar{x}_B}{\sqrt{\dfrac{\sigma_A^2}{n_A} + \dfrac{\sigma_B^2}{n_B}}}$ Use s^2 if σ^2 is not known and the critical value from the z table

Large Sample

Decision

Small Sample

$\sigma_A^2 = \sigma_B^2$ $t = \dfrac{\bar{x}_A - \bar{x}_B}{s_p\sqrt{\dfrac{1}{n_A} + \dfrac{1}{n_B}}}$

with $df = n_a + n_b - 2$

Use F Test

$\sigma_A^2 \neq \sigma_B^2$ $t = \dfrac{\bar{x}_A - \bar{x}_B}{\sqrt{\dfrac{s_A^2}{n_A} + \dfrac{s_B^2}{n_B}}}$

with df the smaller of

$n_A - 1$ or $n_B - 1$

Figure 11–2: Deciding Which Test Statistic To Use For Independent Samples

A Study Guide For Statistics

CHAPTER 11 DISCUSSION QUESTIONS

These questions may be used in your study group or simply as topics for individual reflection. Whichever you do, take time to explain verbally each topic to insure your own understanding. Since these questions are intended as topics for discussion, answers to these questions are not provided. If you find that you are not comfortable with either your answers or that your group has difficulty with the topic, take time to meet with your professor to get help.

1. What is the difference between an independent and dependent sample?

2. How does dependent sampling provide better control of confounding variables?

3. Why must we be concerned with the equality of population variances when we do hypothesis tests with two populations?

4. What is the test statistic for testing $\mu_1 - \mu_2$ for independent samples with equal variances?

5. Why do we use a smaller degree of freedom for a hypothesis tests about $\mu_1 - \mu_2$ when the population variances are not equal?

6. What test statistic do we use if the population variances are unequal and the size of the independent samples are small?

7. When can we use the z test statistic for a hypothesis test with independent samples from two populations?

8. When do we use the pooled estimate of p for a hypothesis test of two population proportions?

9. What is the difference between a controlled experiment and an observational experiment?

10. What is a double-blind randomized controlled experiment?

Comparing Two Treatments

CHAPTER 11 TEST

Given:

	n	$\bar{x}$	s^2
A	21	34.5	10.38
B	30	31.3	4.67

Assume the variances are not equal.

1. What is the value of df for the hypothesis test?

2. What is the value of s_p?

3. State the formula and value of the test statistic for the hypothesis test for $H_0: \mu_A = \mu_B$ and $H_a: \mu_A \neq \mu_B$ at $\alpha = 5\%$.

4. What is your conclusion?

Given: $\bar{d} = 0.572$ $\quad s_d = 1.58$ $\quad n = 10$

5. Test $H_0: \mu_A = \mu_B$ and $H_a: \mu_A \neq \mu_B$ at $\alpha = 5\%$.

6. Write the 95% confidence interval for $\mu_A - \mu_B$.

Given: $n_A = 10$, $\bar{x}_A = 3.15$, $s_A^2 = 1.31$, $n_B = 15$, $\bar{x}_B = 1.87$, $s_B^2 = 1.24$
Assume population variances are equal.

7. What is the value of the appropriate test statistic for the hypothesis test $H_0: \mu_A = \mu_B$ and $H_a: \mu_A > \mu_B$ at $\alpha = 5\%$?

8. Find the 95% confidence interval for $\mu_A - \mu_B$.

Given: $\hat{p}_1 = 0.25$ and $\hat{p}_2 = 0.32$ with $n_1 = 100$ and $n_2 = 100$

9. Test the hypothesis $H_0: p_1 = p_2$ versus $H_a: p_1 \neq p_2$ at $\alpha = 5\%$.

10. Find the 95% confidence interval for $p_1 - p_2$.

CHAPTER 11 TEST Questions and Answers

Given:

	n	$\bar{x}$	s^2
A	21	34.5	10.38
B	30	31.3	4.67

Assume the population variances are equal.

1. What is the value of df for the hypothesis test?

Answer: $df = \dfrac{\left(\dfrac{10.38}{21} + \dfrac{4.67}{30}\right)^2}{\dfrac{\left(\dfrac{10.38}{21}\right)^2}{20} + \dfrac{\left(\dfrac{4.67}{30}\right)^2}{29}} = 34.58 \approx 35$

2. What is the value of s_p?

Answer: $s_p = \sqrt{\dfrac{(n_A - 1)s_A^2 + (n_B - 1)s_B^2}{n_A + n_B - 2}}$

$= \sqrt{\dfrac{20(10.38) + 29(4.67)}{49}}$

$= 2.6459$

Comparing Two Treatments

3. State the formula and value of the test statistic for the hypothesis test for $H_0: \mu_A = \mu_B$ and $H_a: \mu_A \neq \mu_B$ at $\alpha = 5\%$.

Answer: $t = \dfrac{\bar{x}_A - \bar{x}_B}{\sqrt{\dfrac{s_A^2}{n_A} + \dfrac{s_B^2}{n_B}}} = \dfrac{34.5 - 31.3}{\sqrt{\dfrac{10.38}{21} + \dfrac{4.67}{30}}} = \dfrac{3.2}{0.8062} = 3.97$

4. What is your conclusion?

Answer: Reject H_o because $3.97 > t(35, 0.025) = 2.03$

Given: $\bar{d} = 0.572$ $s_d = 1.58$ $n = 10$

5. Test $H_0: \mu_A = \mu_B$ and $H_a: \mu_A \neq \mu_B$ at $\alpha = 5\%$.

Answer: $t = \dfrac{\bar{d}}{\dfrac{s_d}{\sqrt{n}}} = \dfrac{0.572}{\dfrac{1.58}{\sqrt{10}}} = 1.14$

$t(0.025, 9) = 2.26$. Fail to reject H_0.

6. Write the 95% confidence interval for $\mu_A - \mu_B$.

Answer: $\bar{d} - t\left(\tfrac{\alpha}{2}, n-1\right)\dfrac{s_d}{\sqrt{n}} \leq \mu_A - \mu_B \leq \bar{d} + t\left(\tfrac{\alpha}{2}, n-1\right)\dfrac{s_d}{\sqrt{n}}$

$0.572 - 2.26\dfrac{1.58}{\sqrt{10}} \leq \mu_A - \mu_B \leq 0.572 + 2.26\dfrac{1.58}{\sqrt{10}}$

$-0.557 \leq \mu_A - \mu_B \leq 1.701$

Given: $n_A = 10$, $\bar{x}_A = 3.15$, $s_A^2 = 1.31$, $n_B = 15$, $\bar{x}_B = 1.87$, $s_B^2 = 1.24$

7.   What is the value of the appropriate test statistic for the hypothesis test $H_0: \mu_A = \mu_B$ and $H_a: \mu_A > \mu_B$ at $\alpha = 5\%$?

Answer: The variances can be assumed equal because $F = 1.06$ and the critical value for $F = 3.21$.

$$t = \frac{\bar{x}_A - \bar{x}_B}{s_p\sqrt{\dfrac{1}{n_A} + \dfrac{1}{n_B}}} = \frac{3.15 - 1.87}{1.12578} = 1.137$$

$$s_p = \sqrt{\frac{9(1.31) + 14(1.24)}{23}} = 1.12578$$

8.   Find the 95% confidence interval for $\mu_A - \mu_B$.

Answer:

$$(3.15 - 1.87) - 1.96(1.12578) \le \mu_A - \mu_B \le (3.15 - 1.87) + 1.96(1.12578)$$
$$-0.93 \le \mu_A - \mu_B \le 3.49$$

Given: $\hat{p}_1 = 0.25$ and $\hat{p}_2 = 0.32$ with $n_1 = 100$ and $n_2 = 100$

9.   Test the hypothesis $H_0: p_1 = p_2$ versus $H_a: p_1 \ne p_2$ at $\alpha = 5\%$.

Answer: $\hat{p} = \dfrac{x_1 + x_2}{n_1 + n_2} = \dfrac{25 + 32}{200} = 0.285$

$\hat{q} = 1 - \hat{p} = 1 - 0.285 = 0.715$

$$\sigma_{\hat{p}_1 - \hat{p}_2} = \sqrt{\hat{p}\hat{q}\left[\frac{1}{n_1} + \frac{1}{n_2}\right]}$$

Comparing Two Treatments

$$= \sqrt{(0.285)(0.715)[0.01 + 0.01]}$$

$$= \sqrt{0.0040755}$$

$$= 0.0638396$$

$$z = \frac{\hat{p}_1 - \hat{p}_2}{\sqrt{\hat{p}\hat{q}\left[\dfrac{1}{n_1} + \dfrac{1}{n_2}\right]}}$$

$$= \frac{0.25 - 0.32}{0.0638396}$$

$$= -1.096$$

At $\alpha = 0.05$ the critical value of z is ± 1.96. Since the value of the test statistic is **not** in the critical region, we fail to reject the null hypothesis.

10.   Find the 95% confidence interval for $p_1 - p_2$.

Answer:

$$\left(\hat{p}_1 - \hat{p}_2\right) - z\left(\tfrac{\alpha}{2}\right)\sqrt{\frac{\hat{p}_1\hat{q}_1}{n_1} + \frac{\hat{p}_2\hat{q}_2}{n_2}} \leq p_1 - p_2 \leq \left(\hat{p}_1 - \hat{p}_2\right) + z\left(\tfrac{\alpha}{2}\right)\sqrt{\frac{\hat{p}_1\hat{q}_1}{n_1} + \frac{\hat{p}_2\hat{q}_2}{n_2}}$$

$$\tfrac{1}{4} - \tfrac{32}{100} - 1.96\sqrt{\frac{\left(\tfrac{1}{4}\right)\left(\tfrac{3}{4}\right)}{100} + \frac{\left(\tfrac{32}{100}\right)\left(\tfrac{68}{100}\right)}{100}} \leq p_1 - p_2 \leq \tfrac{1}{4} - \tfrac{32}{100} + 1.96\sqrt{\frac{\left(\tfrac{1}{4}\right)\left(\tfrac{3}{4}\right)}{100} + \frac{\left(\tfrac{32}{100}\right)\left(\tfrac{68}{100}\right)}{100}}$$

$$-0.07 - 1.96(0.064) \leq p_1 - p_2 \leq -0.07 + 1.96(0.064)$$

$$-0.07 - 0.125 \leq p_1 - p_2 \leq -0.07 + 0.125$$

$$-0.195 \leq p_1 - p_2 \leq 0.055$$

$$\sum \circ$$

chapter twelve

REGRESSION ANALYSIS–I SIMPLE LINEAR REGRESSION

But of all these principles,
least squares is the most simple:
by the others we would be
led into the most complicated calculations.

– K.F. Gauss (1809)

INTRODUCTION

In this chapter we continue the work we started in chapter 3 as a study of bivariate data. The work is extended to include a discussion of the method of least squares, hypothesis testing, and estimation. As in chapter 3 we are interested in finding a line that fits a set of bivariate data. In general, it is impossible to find a line that all the points will fall on. So we focus on the distances (**errors**) between the points and the line we draw. We want the line that minimizes the squares of the distances using the **least squares method**. This line is called the **line of best fit** or the **regression line**. The equation of the regression line is written as $\hat{y} = \hat{\beta}_0 + \hat{\beta}_1 x$ where $\hat{\beta}_0$ and $\hat{\beta}_1$ are found using formulas that involve the summations we used in earlier chapters; Σx, Σy, Σx^2, Σy^2, and Σxy .

We are then able to use the equation of this line to predict the approximate value of the dependent variable when we know the value of the independent variable. The approximate value we get from the regression line is a point estimate. However, we can also create an interval estimate for the value using a confidence interval.

If few of the points we plot are on the line, we say that the linear relationship is weak with a value of $\hat{\beta}_1$ close to 0. We may then want to test a hypothesis that there is no linear relationship between the two variables. This would be the same as saying that the true value , β_1, of the slope of **the population regression line** is 0. We can again use a t test to test the null hypothesis, H_0: $\beta_1 = 0$ against either a one-tailed or two-tailed alternate hypothesis.

If most of the points we plot are on the line or close to the line, we argue that there is a strong linear relationship between the two variables. We then feel good about using the equation of the regression line to predict values for the dependent variable, $\hat{y}$. The value of $\hat{y}$ we get by substituting a value of x into the equation of the regression line is a point estimate that represents the mean value of y for the value of x. We can also find a confidence interval for $\hat{y}$.

We use what is called the linear **coefficient of determination** to measure the strength of the linear relationship. The coefficient of determination is the fraction of **explained variation** to the **total variation** in y. The **correlation coefficient**, r, is the square root of the coefficient of determination. The correlation coefficient is a value between 0 and 1 for a line with a positive slope, and between -1 and 0 for a line with a negative slope. When all the points fall on the line, $r = 1$ or $r = -1$.

We will continue the work with regression analysis in the next chapter by extending the ideas developed with bivariate data to what is termed multiple linear regression.

A Study Guide For Statistics

CHAPTER 12 HOT SPOTS

1. **The Effect Of Early Rounding Off On The Regression Line And The Correlation Coefficient.**
   Starts on **page 12–4.** J&B 12.4, 12.7

2. **Using A t Test Statistic For A Hypothesis Test Of H_0: $\beta_1 = 0$ Versus H_a: $\beta_1 \neq 0$.**
   Starts on **page 12–6.** Problems on **pages 12–7, 12–8.** J&B 12.6

3. **The Danger Of Using The Regression Line To Predict Values Of y For x Outside The Interval Used To Obtain The Line.**
   Starts on **page 12–8.** J&B 12.6

If you find other HOT SPOTS, write them down and use them as a focus of your discussion in the study group. Or you can use the HOT SPOTS as the topic for a help session with your professor.

Johnson And Bhattacharyya Sections 12.4 And 12.7

Hot Spot #1 – The Effect Of Early Rounding Off On The Regression Line And The Correlation Coefficient

The formulas we use for finding the slope, $\hat{\beta}_1$ and y intercept $\hat{\beta}_0$, of the line of least squares and the correlation coefficient, r, are very sensitive to errors in the summations we use. This is particularly true if the data values are small. The following set of bivariate data will be used to show what happens when errors are made:

x	y	x^2	y^2	xy
1	1.2	1	1.44	1.2
1.5	0.8	2.25	0.64	1.2
2	0.6	4	0.36	1.2
3	0.4	9	0.16	1.2
4	0.3	16	0.09	1.2

Table 12.1: Table of Sample Bi-variate Data

The totals for each column are found as follows:

$$\sum x = 11.5 \quad \sum y = 3.3 \quad \sum x^2 = 32.25 \quad \sum y^2 = 2.69 \quad \sum xy = 6$$

When the above sums are substituted into the formulas for $\hat{\beta}_0$, $\hat{\beta}_1$, and r, we get the following values:

$$\hat{\beta}_1 = -0.27413 \qquad \hat{\beta}_0 = 1.2905 \qquad r = -0.9227$$

We are now ready to see what happens if an error is made in one of the entries in the table. One of the most common mistakes is making an error in squaring one of the values in the table. The values of x and y in the above table are easy to square, but let's see what happens when we make an error in squaring $y = 0.3$. If $(0.3)^2$ were written in error as 0.9, the only sum that would change is Σy^2, which would increase to 3.5. In this case the values of $\hat{\beta}_0$ and $\hat{\beta}_1$ do not change because the formulas do not use Σy^2. But there is a drastic change in the r value. The new value of r is -0.574205.

If we look at the value of r^2 for the two different values of r we obtained, we can really see the effect of the error. The r^2 of 0.8514 for the correct value of r tells us that approximately 85% of the variation in the y values is due to the regression line. When the error is made, the new r^2 is approximately 33%, which is a reduction of more than 50% in explained variation.

Now let's look at errors that also affect values of $\hat{\beta}_0$ and $\hat{\beta}_1$. If the error were made in a value of x^2, the error changes the value of the denominator in the formula for both $\hat{\beta}_1$ and r and thus changes the value of all three computations. We can illustrate this last point by looking at what happens if the square of 1.5 is written incorrectly as 22.5 rather than 2.25. The Σx^2 is increased by 20.25 and the effect on $\hat{\beta}_0$, $\hat{\beta}_1$, and r is again drastic. The new values based on the error are as follows:

$$\hat{\beta}_1 = -0.0610 \qquad \hat{\beta}_0 = 0.8004 \qquad r = -0.4354$$

Of the two errors we examined this last one is the more serious because all three values are affected to such a degree that we might as well throw out the entire analysis. Again the point is that care must be taken in the calculations used in the formulas for $\hat{\beta}_0$, $\hat{\beta}_1$, and r.

The obvious solution to this problem of errors is to use a calculator or computer to obtain the values for the regression line and the correlation coefficient. If we must do the calculations using the formulas, it is imperative that we avoid errors in the table of values we create to find $\hat{\beta}_0$, $\hat{\beta}_1$, and r.

Johnson And Bhattacharyya Section 12.6

Hot Spot #2 – Using A t Test Statistic For A Hypothesis Test Of H_0: $\beta_1 = 0$ Versus H_a: $\beta_1 \neq 0$

The hypothesis test for the null hypothesis H_0: $\beta_1 = 0$ fits the template model used in chapter 9 (page 9–11) where

$$z = \frac{[] - \mu_{[]}}{\sigma_{[]}}, \text{ where the [] can be replaced by}$$

1.	x
2.	$\bar{x}$
3.	$\hat{p}$
4.	$\bar{x}_2 - \bar{x}_1$
5.	$\bar{d}$

We can now add the sample statistic $\hat{\beta}_1$ to our list:

6. $\hat{\beta}_1$

In the case of $\hat{\beta}_1$, when $\beta_1 = 0$, the sampling distribution for small samples with $n \leq 30$ involves a t test with $n - 2$ degrees of freedom. We lose the 2 degrees of freedom because we need to calculate both $\bar{x}$ and $\bar{y}$.

Returning to our template, we need to replace z with t. When we place $\hat{\beta}_1$ in the box, the t test looks like the following:

$$t = \frac{\hat{\beta}_1 - \mu_{\hat{\beta}_1}}{\sigma_{\hat{\beta}_1}} \text{ where } \mu_{\hat{\beta}_1} = \beta_1 \text{ and } \sigma_{\hat{\beta}_1} = \frac{\sigma}{\sqrt{SS_{xx}}}$$

The final form of our test statistic is:

$$t = \frac{\hat{\beta}_1 - \beta_1}{s_{\hat{\beta}_1}} \text{ with } df = n - 2$$

Returning to the hypothesis test of $\beta_1 = 0$, once we have a value of $\hat{\beta}_1$ and s the t test is easy to do. We follow the same five steps we have been using for hypothesis testing in the last three chapters.

      Step 1. State the hypotheses.

      Step 2. State the test statistic to use.

      Step 3. Determine the rejection region.

      Step 4. Find the value of the test statistic.

      Step 5. Either reject or fail to reject the null hypothesis.

Hot Spot #2 Sample Problem and Solution: Given $\hat{\beta}_1 = 0.3, s_{\hat{\beta}_1} = 0.225$, and $n = 20$, test $H_0: \beta_1 = 0$ and $H_a: \beta_1 \neq 0$ at the 5% level of significance.

Answer: Step 1. $H_0: \beta_1 = 0$ and $H_a: \beta_1 \neq 0$.

           Step 2. $n \leq 30$ and $s_{\hat{\beta}_1}$ is used for $\sigma_{\hat{\beta}_1}$, so we use t.

           Step 3. The critical value of t for a two-tailed test at $df = 18$ is

                $t = \pm 2.101$.

           Step 4. $t = \dfrac{0.3}{0.225} = 1.33$

           Step 5. Fail to reject.

Sample Problem and Answer: Given $\hat{\beta}_1 = 0.4$, $s_{\hat{\beta}_1} = 0.173$, and $n = 29$ test $H_0: \beta_1 = 0$ and $H_a: \beta_1 > 0$ for significance at the 5% level.

Answer: Step 1. $H_0: \beta_1 = 0$ and $H_a: \beta_1 > 0$.

           Step 2. $n \leq 30$ and $s_{\hat{\beta}_1}$ is used for $\sigma_{\hat{\beta}_1}$, so we use t.

           Step 3. The critical value of t for a one-tailed test at $df = 27$ is

                $t = 1.703$.

Step 4. $t = \dfrac{0.4}{0.173} = 2.31$

Step 5.   Reject H_0 in favor of the alternative H_a: $\beta_1 > 0$.

Sample Problem and Answer: Given $\hat{\beta}_1 = 0.2$, $s_{\hat{\beta}_1} = 0.099$, and $n = 100$ test H_0: $\beta_1 = 0$ and H_a: $\beta_1 \neq 0$. for significance at the 5% level.

Answer: Step 1. H_0: $\beta_1 = 0$ and H_a: $\beta_1 \neq 0$..

Step 2. $n > 30$, so we use the normal distribution for the critical value t.

Step 3.   The critical value of t for a two-tailed test at $df = 98$ is $t = \pm 1.96$.

Step 4. $t = \dfrac{0.2}{0.099} = 2.02$

Step 5.   Reject H_0 in favor of the alternative H_a: $\beta_1 \neq 0$.

✳
*Johnson And
Bhattacharyya
Section 12.6*

Hot Spot #3– The Danger Of Using The Regression Line To Predict Values Of y For x Outside The Interval Used To Obtain The Line

Many times the scatter diagrams appear to be clustered around a line when in fact the relationship between the variables is curvilinear. The apparent linear relationship may be because the values of x are restricted to a particular interval. The equation $xy = 12$, for x and y both positive, gives a good illustration of the problems that can exist. The graph of the equation is a hyperbola in the first quadrant as shown below, in Figure 12–9:

A Study Guide For Statistics

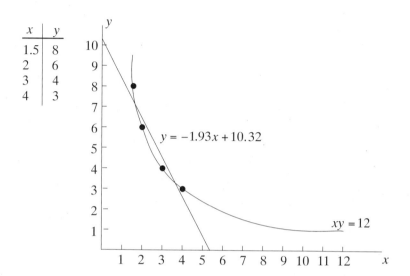

x	y
1.5	8
2	6
3	4
4	3

$y = -1.93x + 10.32$

$xy = 12$

Figure 12–1:
A Scatter
Diagram Of
Values On The
Curvi-Linear
Hyperbola
$xy=12$

If the bivariate set of data is limited to values of x between 1 and 4 as illustrated in the above data set, the graph appears to cluster around the regression line $y = -1.932x + 10.322$ that is found using the 4 ordered pairs in the table. If we use the regression line to find the value of y when x is 10, we get

$$y = -1.932(10) + 10.322 = -9$$

It is clear from the graph of $xy = 12$ that the y value must always be positive. The correct value for y is found by substituting 10 into the equation $xy = 12$.

$$y = \frac{12}{10} = 1.2$$

Yet when we use the regression line we get a negative value of y when $x = 10$. The residual e is

$$e = 1.2 - (-9) = 10.2$$

The error is 850% $\left(\frac{10.2}{1.2}\right)$. The problem is that the linear relationship we created with the regression line was only valid for the interval from 1 through 4. When we use the regression line to find y for $x = 10$ we are assuming that the linear relationship is still valid. In this example we made a large error in saying $y = -9$ when $x = 10$. As a rule of thumb, we should stay within 20% of the width of our original interval. In the above example the width of the original interval is 3 and 20% of 3 is 0.6.

We can now see how close we are to the correct value of y when we use the largest value of x allowed, that is, $4 + (0.02)(3) = 4.6$:

$$y = -1.932(4.6) + 10.322 = 1.435$$

The correct value of y is found by substituting 4.6 into the equation $xy = 12$.

$$y = \tfrac{12}{4.6} = 2.609$$

We still have an error, in that the residual value $\hat{e}$ is

$$\hat{e} = 1.435 - 2.609 = -1.174$$

The error of 45% $\left(\tfrac{1.174}{2.609}\right)$ is large; but when compared to the error of 850% that we had using $x = 10$, we have a considerable improvement. The point is that we will have an error in either case. But if we limit ourselves to 20% of the width of the original interval for x, the error will at least be reasonable.

Of course, in practice we never know if the bivariate data set we use for linear regression actually belongs to a curvilinear relation. Relative to the interval $(1,4)$ we used to get the regression line, the correlation coefficient $r = 0.966$. When we square r, we get the coefficient of determination r^2. This value is 0.933 and it tells us that 93% of the variation in y is explained by the regression line. The point is that relative to our interval of $(1,4)$ the regression line is a good model that explains the clustering we have about the line. It is only when we go outside of the interval that we have the potential for making large errors in estimating y. Of course, if our correlation coefficient is very small we may still make large errors in estimating y independent of whether or not there is a curvilinear relationship. Indeed it is possible to have bivariate data sets that exhibit no relationship whatsoever.

CHAPTER 12 DISCUSSION QUESTIONS

These questions may be used in your study group or simply as topics for individual reflection. Whichever you do, take time to explain verbally each topic to insure your own understanding. Since these questions are intended as topics for discussion, answers to these questions are not provided. If you find that you are not comfortable with either your answers or that your group has difficulty with the topic, take time to meet with your professor to get help.

1. What is the purpose of using a transformation of the data set in bivariate data analysis?

2. What is a residual in bivariate data analysis?

3. How is the coefficient of determination used to evaluate the degree of linear relationship between two variables?

4. What is the t test statistic that is used with hypothesis testing of H_0: $\beta_1 = 0$ versus H_a: $\beta_1 \neq 0$?

5. Why does the interval estimate of y change as the values of x are further away from the mean of x?

6. Why is a linear relationship between two variables not necessarily a causal relationship?

7. How is it possible that a weak linear relationship between two variables can result in a significant value of $\hat{\beta}_1$ in a hypothesis test of H_0: $\beta_1 = 0$ versus H_a: $\beta_1 \neq 0$?

8. How does early rounding off of values affect both regression and correlation analysis?

9. What are the assumptions of the linear regression model?

10. How might it be possible that two variables are related even though the linear correlation coefficient is nearly zero?

CHAPTER 12 TEST

Use the following information for problems 1, 2 and 3:

$$\sum x = 10 \qquad \sum y = 14 \qquad \sum xy = 58 \qquad \sum x^2 = 44 \qquad \sum y^2 = 80 \qquad n = 4$$

1. Find b and a.

2. Find r.

3. Find r^2.

4. Given $b = 0.3$, $s_b = 0.138$, and $n = 50$, use a t test to do a hypothesis test on $H_o:B = 0$ versus $H_a:B \neq 0$ at $\alpha = 0.05$.

5. Given $\sum (y - \bar{y})^2 = 122.5$ and $\sum e^2 = 43.2$, find r^2.

6. Given the regression line $\hat{y} = 1.25x + 4.5$, find a point estimate for y when $x = 1.5$.

7. Given that the ordered pair $(1.5, 6.1)$ belongs to the bivariate data set, find the error for the estimate in problem 6

8. Find the 95% confidence interval for the mean value of y when $x = 1.5$ in problem 6.

Use the following information for problems 9 and 10:

$$SST = 0.235 \qquad SSE = 0.025 \qquad SSR = 0.21 \qquad n = 10$$

9. Calculate s_e.

10. Calculate r^2.

CHAPTER 12 TEST Questions and Answers

Use the following information for problems 1, 2 and 3:

$$\sum x = 10 \qquad \sum y = 14 \qquad \sum xy = 58 \qquad \sum x^2 = 44 \qquad \sum y^2 = 80 \qquad n = 4$$

1. Find b and a.

Answer: $b = \dfrac{\left(\sum xy\right) - \dfrac{\left(\sum x\right)\left(\sum y\right)}{n}}{\left(\sum x^2\right) - \dfrac{\left(\sum x\right)^2}{n}}$

$$b = \frac{58 - \dfrac{(10)(14)}{4}}{44 - \dfrac{(10)^2}{4}} = \frac{92}{76} = 1.2105$$

$$\bar{x} = 2.5 \qquad \bar{y} = 3.5$$

$$a = \bar{y} - b\bar{x} = 3.5 - 3.026 = 0.4737$$

2. Find r.

Answer: $r = \dfrac{SS_{xy}}{\sqrt{SS_{xx}}\sqrt{SS_{yy}}} = \dfrac{\left(\sum xy\right) - \dfrac{\left(\sum x\right)\left(\sum y\right)}{n}}{\sqrt{\left(\sum x^2\right) - \dfrac{\left(\sum x\right)^2}{n}}\sqrt{\left(\sum y^2\right) - \dfrac{\left(\sum y\right)^2}{n}}}$

$$r = \frac{(58) - \dfrac{(10)(14)}{4}}{\sqrt{(44) - \dfrac{(10)^2}{4}}\sqrt{(80) - \dfrac{(14)^2}{4}}} = \frac{23}{\sqrt{19}\sqrt{31}} = 0.9477$$

3. Find r^2.

Answer: $r^2 = (0.9477)^2 = 0.8981$

4. Given $b = 0.3$, $s_b = 0.138$, and $n = 50$ use a t test to do a hypothesis test on $H_o : B = 0$ versus $H_a : B \neq 0$ at $\alpha = 0.05$.

Answer: $df = 48$ $t_{cv} = \pm 1.96$ $t = \dfrac{0.3}{0.138} = 2.18$

Conclusion: Reject H_0.

5. Given $\sum (y - \bar{y})^2 = 122.5$ and $\sum e^2 = 43.2$, find r^2.

Answer: $SSR = SST - SSE = 122.5 - 43.2 = 79.3$

$$r^2 = \frac{SSR}{SST} = \frac{79.3}{122.5} = 0.65$$

6. Given the following graph of $y = a + bx$, estimate b and a:

Answer:

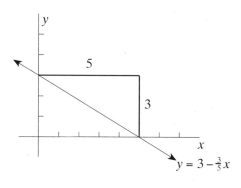

$$a = 3 \text{ and } b = -\tfrac{3}{5}$$

7. Given the regression line $\hat{y} = 1.25x + 4.5$, find a point estimate for y when $x = 1.5$.

Answer: $\hat{y} = 1.25(1.5) + 4.5 = 1.875 + 4.5 = 6.375.$

8. Given that the ordered pair $(1.5, 6.1)$ belongs to the bivariate data set, find the error for the estimate in problem 7.

Answer: $e = \hat{y} - y = 6.375 - 6.1 = 0.275.$

Use the following information for problems 9 and 10:

$SST = 0.235 \qquad SSE = 0.025 \qquad SSR = 0.21 \qquad n = 10$

9. Calculate s_e.

Answer: $s_e = \sqrt{\dfrac{SSE}{n-2}} = \sqrt{\dfrac{0.025}{8}} = \sqrt{0.003125} = 0.055901699$.

10. Calculate r^2.

Answer: $r^2 = \dfrac{SSR}{SST} = \dfrac{0.21}{0.235} = 0.8936$.

$\sum\circ$

chapter thirteen

REGRESSION ANALYSIS–II MULTIPLE LINEAR REGRESSION AND OTHER TOPICS

Suit the action to the word,
the word to the action,
with this special observance,
that you o'erstep not the modesty of nature . . .
— HAMLET Act 3, Scene 2

INTRODUCTION

Our work in the last chapter involved simple linear regression. In this chapter we extend that work to **multiple linear regression** where we have two or more independent variables. Many of the concepts that were developed in simple regression can, with minimal effort, be extended to multiple linear regression analysis involving k independent variables, $x_1, x_2, ..., x_k$.

The idea of a **residual** is still written as $\hat{e}_i = \hat{y}_i - y_i$, and we are still interested in the distribution of $\hat{e}_i$. The hypothesis test of H_0: $\beta_0 = 0$ with

$df = n - 2$ is extended to a more general form $H_0: \beta_i = 0$ where the degrees of freedom equal $n - k - 2$ with k independent variables. We again use a t test to test the null hypothesis, $H_0: \beta_i = 0$ against a one or two-tailed alternate hypothesis.

Whether it is one independent variable or many independent variables, we are still interested in trying to explain the variability in the dependent variable ,y, in terms of the relation defined in our model. The calculations that we did in the last chapter using formulas to find the coefficient of the independent variable and the correlation coefficient are done in multiple regression using a computer. As a result, more of our time can be spent in analyzing the computer print out that focuses on

1. the estimated regression coefficients, $\hat{\beta}_i$,

2. the corresponding estimated standard errors and the related value of the t test statistic for each $\hat{\beta}_i$,

3. the **multiple correlation coefficient,** R^2,

4. the standard deviation of y about the regression line, and

5. the distribution of $\hat{e}_i$.

Other topics covered in this chapter relate to **nonlinear regression analysis** involving bivariate data and what can be done to transform both dependent and independent variables to insure that the **transformed data set** shows a linear relation.

The chapter finishes with a general look at **statistical models** and what we should do to examine the adequacy of the statistical model. This is done using a plot of the residuals $\hat{e}_i$ versus the predicted value $\hat{y}_i$ to check the assumption of constant error variance and a plot of the residuals $\hat{e}_i$ versus time order to determine if the errors e_i are independent.

CHAPTER 13 HOT SPOTS

1. **The Effect Of Transformations On The Regression Line And The Correlation Coefficient.**

   Starts on **page 13–4.** Problems on **pages 13–6, 13–7.** J&B 13.2

If you find other HOT SPOTS, write them down and use them as a focus of your discussion in the study group. Or you can use the HOT SPOTS as the topic for a help session with your professor.

Johnson And
Bhattacharyya
Section 13.2

Hot Spot #1– The Effect Of Transformations On The Regression Line And The Correlation Coefficient

Many times the scatter diagrams appear to be clustered around a line when in fact the relationship between the variables is curvilinear. The apparent linear relationship may be because the values of x are restricted to a particular interval. The equation $xy = 12$, for x and y both positive, gives a good illustration of the problems that can exist. The graph of the equation is a hyperbola in the first quadrant as shown below, in Figure 13–1:

Figure 13–1: A Scatter Diagram Of Values On The Curvi-Linear Hyperbola xy=12

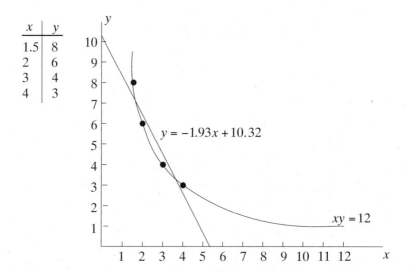

x	y
1.5	8
2	6
3	4
4	3

$y = -1.93x + 10.32$

$xy = 12$

If the bivariate set of data is limited to values of x between 1 and 4 as illustrated in the above data set, the graph appears to cluster around the regression line $y = -1.932x + 10.322$. Using the 4 ordered pairs in the table, the correlation coefficient $r = 0.966$. When we square r, we get the coefficient of determination r^2. This value is 0.933 and it tells us that 93% of the variation in y is explained by the regression line.

The point is that relative to our interval of $(1, 4)$ the regression line is a reasonably good model that explains the clustering we have about the line. It is only when we go outside of the interval that we have the potential for making a serious error in predicting the value of y. For example, if we use values of $x > 5.5$, the predicted value of y will be negative. The original relationship defined by the equation $xy = 12$ results in $y > 0$ for all values of x. Of course, in practice we never know if the bivariate data set we use for linear regression actually belongs to a curvilinear relation.

Often times, we can avoid the difficulties presented in the above problem by finding a suitable transformation of one of the variables. The usual transformations used involve taking logs, reciprocals, or square roots of the variables. The choice of the transformation to use is often suggested by the scatter diagram. In the above example the data suggests that the relationship between the variables involves the equation $y = \dfrac{1}{x}$. As such, we are interested in the transformation $y' = \dfrac{1}{y}$ and $x' = x$. In other situations, the desired transformation may be determined from information known about the data set itself.

We will use the reciprocal transformation $y' = \dfrac{1}{y}$ and $x' = x$ to get a transformed model $y' = \beta_0 + \beta_1 x'$. The new table of ordered pairs is:

x'	1.5	2	3	4
y'	$\frac{1}{8}$	$\frac{1}{6}$	$\frac{1}{4}$	$\frac{1}{3}$

Using the HP-21S to analyze the transformed data set, the values of the estimated coefficients and the correlation coefficient are as follows

$$\hat{\beta}_0 = -2.2033 \times 10^{-11} \quad \hat{\beta}_1 = 0.08333333 \quad r = 1$$

The equation of the regression line for the transformed model is

$$y' = 0.08333333x'.$$

Clearly, we have a much better model that explains 100% of the variability of the dependent variable. Furthermore, if we go outside the original interval, we will obtain an accurate estimate for y using $y = \dfrac{1}{y'}$.

Hot Spot #1 Sample Problem and Solution: Given the following data set, plot the scatter diagram, determine the appropriate transformation, and find the equation of the regression line for the transformed model.

x	$\frac{3}{4}$	1	$\frac{5}{4}$	$\frac{7}{4}$
y	2.1	2.7	3.5	5.8

Answer: Step 1: The scatter diagram

Step 2. The scatter diagram appears to be exponential and suggests that we should use the following transformation:

$$y' = \log_e y \text{ and } x' = x$$

Step 3: The transformed data set is as follows:

x'	$\frac{3}{4}$	1	$\frac{5}{4}$	$\frac{7}{4}$
y'	0.74	0.99	1.25	1.75

Step 4:   The equation of the regression line for the transformed model is $y' = 1.01x'$.

Sample Problem and Answer: Given the following data set, determine the appropriate transformation, the transformed data set, and find the equation of the regression line for the transformed model

x	1	2	3	4
y	1	3.8	9.2	15.7

Answer: Step 1.   The scatter diagram and the data set suggest the relationship is defined by the equation $y = x^2$. The appropriate transformation is $y' = \log y$ and $x' = \log x$.

Step 2.   The transformed data set is as follows:

x'	0	0.30	0.48	0.60
y'	0	0.58	0.96	1.20

Step 3.   The equation of the regression line for the transformed model is $y' = 2x' - 0.0065$.

Sample Problem and Answer: Given the following data set, determine the appropriate transformation.

x	1	4	6	10
y	1.1	1.8	2.5	3.2

Answer: The data suggests using $y' = y$ and $x' = \sqrt{x}$.

CHAPTER 13 DISCUSSION QUESTIONS

These questions may be used in your study group or simply as topics for individual reflection. Whichever you do, take time to explain verbally each topic to insure your own understanding. Since these questions are intended as topics for discussion, answers to these questions are not provided. If you find that you are not comfortable with either your answers or that your group has difficulty with the topic, take time to meet with your professor to get help.

1. What is the purpose of transforming either the x or y values in a bivariate data analysis?

2. How does one check for a normal distribution of residuals?

3. How does one check for constant error variance?

4. What is the t test statistic that is used with hypothesis testing of H_0: $\beta_i = 0$ versus H_a: $\beta_i \neq 0$?

5. What does one do if the assumptions of the multiple linear regression model are violated?

6. What is checked with a plot of residuals versus time order?

7. What is R^2?

8. How does early rounding off of values affect both regression and correlation analysis?

9. What are the assumptions of the multiple linear regression model?

10. What is the difference between a polynomial regression model and a multiple regression model?

A Study Guide For Statistics

CHAPTER 13 TEST

Use the following information for problems 1, 2, 3, and 4:

x	1	4	6	10
y	1.1	1.8	2.5	3.2

1. Determine the linearizing transformation.

2. Find the transformed data set.

3. Find the equation of the regression line for the transformed data.

4. What proportion of variability is explained by the fitted model?

A multiple linear regression was fitted to a data set obtained from 11 runs of an experiment in which two predictors, x_1 and x_2, were observed along with the response y. The following results were obtained:

$$\hat{\beta}_0 = -0.3 \qquad \hat{\beta}_1 = 2.16 \qquad \hat{\beta}_2 = 19.4$$

$$SS\text{ due to regression} = 732.4 \qquad\qquad SSE = 56.27$$

5. State the degrees of freedom.

6. Predict the response variable for $x_1 = 12$ and $x_2 = 0.3$.

7. Estimate the error standard deviation σ.

8. What proportion of the y-variability is explained by the fitted regression?

The estimated standard errors of $\hat{\beta}_1$ and $\hat{\beta}_2$ were 0.06 and 2.47 respectively.

9. Obtain a 90% confidence interval for β_2.

10. Test $H_0: \beta_1 = 3$ versus $H_1: \beta_1 < 3$ with $\alpha = 0.05$.

Regression Analysis–II (Multiple Linear Regression and Other Topics)

CHAPTER 13 TEST Questions and Answers

Use the following information for problems 1, 2, 3, and 4:

x	1	4	6	10
y	1.1	1.8	2.5	3.2

1.   Determine the linearizing transformation.

Answer: Use $y' = y$ and $x' = \sqrt{x}$.

2.   Find the transformed data set.

Answer:

x'	1	2	2.45	3.16
y'	1.1	1.8	2.5	3.2

3.   Find the equation of the regression line for the transformed data.

Answer: $y' = 0.988x' + 0.0234$

4.   What proportion of variability is explained by the fitted model?

Answer: $r^2 = (0.989)^2 = 0.978$

A multiple linear regression was fitted to a data set obtained from 11 runs of an experiment in which two predictors, x_1 and x_2, were observed along with the response y. The following results were obtained:

$$\hat{\beta}_0 = -0.3 \qquad \hat{\beta}_1 = 2.16 \qquad \hat{\beta}_2 = 19.4$$

$$SS \text{ due to regression} = 732.4 \qquad\qquad SSE = 56.27$$

5. State the degrees of freedom.

Answer: $n - k - 1 = 11 - 2 - 1 = 8$

6. Predict the response variable for $x_1 = 12$ and $x_2 = 0.3$.

Answer: $\hat{y} = -0.3 + 2.16(12) + 19.4(0.3) = 31.44$

7. Estimate the error standard deviation σ.

Answer: $s = \sqrt{\dfrac{SSE}{n-3}} = \sqrt{\dfrac{56.27}{8}} = 2.66$

8. What proportion of the y-variability is explained by the fitted regression?

Answer: $r^2 = \dfrac{SS \text{ due to regression}}{\text{Total SS of } y} = \dfrac{732.40}{788.67} = 0.93$

The estimated standard errors of $\hat{\beta}_1$ and $\hat{\beta}_2$ were 0.6 and 2.47 respectively.

9.   Obtain a 90% confidence interval for β_2.

Answer: The 90% confidence interval for β_2 is given by

Estimated coefficient $\pm t_{\alpha/2}$ (estimated S.E.)

$19.4 \pm 2.306(2.47)$ or $(13.704,\ 25.096)$

10.   Test $H_0:\beta_1 = 3$ versus $H_1:\beta_1 < 3$ with $\alpha = 0.05$.

Answer: $t = \dfrac{\hat{\beta}_1 - \beta_1}{\text{Estimated S.E.} \left(\hat{\beta}_1\right)} = \dfrac{2.16 - 3}{0.6} = -1.4$ and $t_{0.05} = -1.860$.

Fail to reject the null hypothesis with $\alpha = 0.05$.

$$\sum{}^{\circ}$$

chapter fourteen

ANALYSIS OF
CATEGORICAL DATA

Don't ask what it means,
but rather how it is used.

— L. Wittgenstein

INTRODUCTION

This chapter is an extension of the hypothesis testing we did with
two proportions. We will use the Chi-Square distribution to do
hypothesis tests that involve more than two population proportions.

The test statistic used in this chapter involves what are called
observed frequencies and **expected frequencies** that occur when we
work with **categorical data**. A **contingency table** is used to record the
observed frequencies in the **cells** of the table. The value of the test
statistic is found by taking the sum of the squares of the difference
between the observed and expected frequencies divided by the expected
frequency for each category. The test statistic has an approximate chi-
square distribution, and the critical value for a given level of significance
is obtained from the chi-square table. The three hypothesis tests that we
will look at in this chapter are called a test for goodness of fit, a test of
independence, and test for homogeneity.

The **test for goodness of fit** is one of the first hypothesis tests developed; it was first published in 1900 by Karl Pearson. This test can be illustrated using an example with coins. In tossing a fair coin, there are two categories; heads and tails. In ten tosses, we expect 5 heads and 5 tails, but we may only observe 4 heads and 6 tails. A hypothesis test of the equality of the proportions for heads and tails is done using the expected values of 5 and the observed values of 4 and 6 in the test statistic. We then compare it to a critical value obtained from the χ^2 table. Of course, the goodness of fit test can be extended to more categories.

The **test of independence** is used to determine if two or more characteristics of a single population are related. A typical example of a test of independence involves data on cholesterol levels for men and women to determine if cholesterol levels are independent of gender. The notation used to find the expected proportion in each cell is similar to what we used in the probability chapter when we had independent events. For example, the expected proportion of the sample that is female and has low cholesterol, p_{FL}, is found using the product $p_F \cdot p_L$. The observed and expected frequencies are then used in the test statistic, and it is compared to a critical value from the χ^2 table.

The **homogeneity test** is similar to the test of independence, but it involves more than one population. For example, we consider five different ethnic groups and measure the number of people in each group with high cholesterol. We test the hypothesis that the five population proportions are equal. The equality of the proportions is used to obtain the expected value for each category. The observed and expected frequencies are then used to get the value of the chi-square test statistic, and it is compared to the critical value from the χ^2 table.

Although the chi-square test statistic that is used in each of the above tests looks very different from the other test statistics we have used in hypothesis testing, we still use the same five steps to determine if we should reject the null hypothesis.

Chapter 14 Hot Spots

1. *z* **Test For Proportions Versus Chi-Square Test.**

   Starts on **page 14–4.** J&B 14.3

2. **Reporting** *P* **Values For The** χ^2 **Distribution**

   Starts on **page 14–6**. Problems on **page 14-8.**

   J&B 14.2, 14.3, 14.4

If you find other HOT SPOTS, write them down and use them as a focus of your discussions in the study group. Or you can use the HOT SPOT as the topic for a help session with your professor.

❋
Johnson And
Bhattacharyya
Section 14.3

Hot Spot #1 – z Test For Proportions Versus Chi-Square Test

Earlier we used a z test statistic to test a hypothesis of equal proportions. The chi-square test of homogeneity can also be used to test a hypothesis that two population proportions are equal. The chi-square homogeneity test has the advantage of being a more general hypothesis test that allows us to test whether two or more populations are homogeneous with respect to a characteristic. The following example shows that we get exactly the same results using the two different hypothesis tests on the same data. A survey of 50 men and 50 women asked them whether they liked country music. The results are entered into the table below:

Figure 14–1:
Survey Of
Country Music
Tastes

Survey of Country Music Tastes			
Respondents	No	Yes	Total
Men(1)	20	30	50
Women(2)	15	35	50
Totals	35	65	100

We want to test a null hypothesis that the population proportions are equal, against the alternate hypothesis that they are not equal, at $\alpha = 5\%$.

$$H_0: p_1 = p_2 \text{ versus } H_a: p_1 \neq p_2$$

The critical values for the two-tailed test are ±1.96, and we will test the hypothesis using the test statistic

$$z = \frac{\hat{p}_1 - \hat{p}_2}{\sqrt{\hat{p}\hat{q}\left[\frac{1}{n_1} + \frac{1}{n_2}\right]}}$$

$$\hat{p}_1 = \frac{30}{50} \qquad\qquad \hat{p}_2 = \frac{35}{50}$$

$$\hat{p} = \frac{30+35}{100} = \frac{65}{100} \qquad\qquad \hat{q} = \frac{35}{100}$$

$$z = \frac{\frac{30}{50} - \frac{35}{50}}{\sqrt{\frac{65}{100}\left(\frac{35}{100}\right)}\sqrt{\frac{1}{50} + \frac{1}{50}}} = -1.048$$

The value of z is between the critical values of -1.96 and 1.96; hence, we fail to reject the null hypothesis. Using the HP-21S, the reported P value is 0.147, which we must double because of the two-tailed test. The value would be at 29.4%, which is much too high.

Now we will do the hypothesis test using the chi-square test of homogeneity. We first need to find the expected values for each cell of the table, and then calculate X^2.

$$E_{1,1} = \frac{50(35)}{100} = 17.5 \qquad E_{1,2} = \frac{50(65)}{100} = 32.5$$

$$E_{2,1} = \frac{50(35)}{100} = 17.5 \qquad E_{2,2} = \frac{50(65)}{100} = 32.5$$

$$X^2 = \frac{(20-17.5)^2}{17.5} + \frac{(30-32.5)^2}{32.5} + \frac{(15-17.5)^2}{17.5} + \frac{(35-32.5)^2}{32.5}$$

$$X^2 = 0.357 + 0.192 + 0.357 + 0.192 = 1.098$$

The critical value of χ^2 at one degree of freedom is 3.84. The value of X^2 is less than the critical value; hence, we again fail to reject the null hypothesis. Using the HP-21S the reported P value for α is 29.4%. This is exactly the same value we found using the z test.

There are two additional points that need to be addressed. The first is that the chi-square test for homogeneity is a two tailed hypothesis test; whereas the z test used with a null hypothesis of equal proportions can be used with a one sided alternative hypothesis. The second point is that the chi-square critical value in the above example was the square of the critical value of z. In general

$$z^2{}_{\alpha/2} = \chi_\alpha{}^2 \text{ at df} = 1.$$

Although the chi-square test is more general, only the z test can be used for a one tailed hypothesis test.

Johnson And
Bhattacharyya
Sections 14.2,
14.3, and 14.4

Hot Spot #2– Reporting P Values For The χ^2 Distribution

The table used to get critical values for the chi-square distribution is similar to the t table. The critical values listed for each value of α are found by first finding the degrees of freedom. However, there is a very important differences in how we use the χ^2 table when we find the critical value for the test of hypothesis for contingency tables.

The null hypothesis used with contingency tables is always a statement of equality. The following table shows the null hypothesis in each of the hypothesis tests involved with contingency tables:

Goodness of Fit Test	$H_0: p_1 = p_2 =,...,= p_k$
Test of Homogeneity	H_0: In each response category, the probabilities are equal.
Test of Independence	H_0: Each cell probability equals the product of the corresponding row and column marginal probabilities.

Even though the alternate hypothesis in each test is two sided, we use the critical value of χ^2 for the right side of the curve because only a large value of the test statistic indicates that we should reject H_0. In contrast, a two sided hypothesis test using the t distribution uses the $\pm$ symbol with the critical value on the right side of the curve to get the two critical values for a two sided hypothesis test.

The chi-square distribution unlike the z and t distributions, is not symmetrical. Because of the non-symmetry, the critical values are listed for each of the two tails of the curve. When we look at the chi-square table we find a heading that gives areas to the right of the critical value for both sides of the curve. A typical heading and row of critical values for $df = (2-1)(2-1) = 1$ follows:

df	$\chi^2(0.99)$	$\chi^2(0.975)$	$\chi^2(0.95)$	$\chi^2(0.50)$	$\chi^2(0.10)$	$\chi^2(0.05)$	$\chi^2(0.01)$
1	0.0002	0.001	0.004	0.45	2.71	3.84	6.63

A Study Guide For Statistics

For example, at $\alpha = 0.05$ for $df = 1$ the critical value of χ^2 for a hypothesis test using a contingency table would use the right side entry of 3.84. We would not use value from the left side of the curve. We would reject H_0 if the observed value of our test statistic was greater than 3.84. As was the case with the test of hypotheses we did with the t distribution, we can use a reported P value approach. But we run into the same difficulty we had with reporting P values with the t distribution.

Before we look at how we handle the problem with the chi-square table, it is again important to understand that we can avoid the problem using a calculator like the HP-21S. Once we have the degrees of freedom entered into the calculator, we can find the area to the right of our observed value by pressing one key. For that matter, we can get critical values for a hypothesis test just as easily. However, the remainder of the discussion assumes that we will use the traditional χ^2 table to report the P value associated with a hypothesis test.

As an example, we will use the illustration from Hot Spot #1. We want to test a null hypothesis that the population proportions are equal, against the alternate hypothesis that they are not equal, at $\alpha = 0.05$ for $df = 1$. We state the hypothesis test as follows:

$$H_0: p_1 = p_2 \text{ versus } H_a: p_1 \neq p_2$$

We compare the calculated value to the table entries for $df = 1$.

$$X^2 = \frac{(20 - 17.5)^2}{17.5} + \frac{(30 - 32.5)^2}{32.5} + \frac{(15 - 17.5)^2}{17.5} + \frac{(35 - 32.5)^2}{32.5}$$

$$X^2 = 0.357 + 0.192 + 0.357 + 0.192 = 1.098$$

The critical value of χ^2 at one degree of freedom is 3.84. The value of X^2 is less than the critical value; hence, we again fail to reject the null hypothesis. Using the HP-21S, the reported P value for α is 29.4%.

Analysis Of Categorical Data

The calculated value was between 0.45 at $\alpha = 50\%$ and 2.71 at $\alpha = 10\%$. We would say that the hypothesis test was significant at a value between 50% and 10%. (Using the HP-21S, the P value = 29.4%)

Hot Spot #2 Sample Problem and Solution: Find the reported P value for the following hypothesis Test of Independence with $X^2 = 7.03$ and $df = 3$.

Solution: $X^2 = 7.03$ is between 6.25 at $\alpha = 0.10$ and 7.81 at $\alpha = 0.05$. The reported value of P is between 5% and 10%.

Sample Problem and Answer: Find the reported P value for a hypothesis test of Goodness of Fit with $X^2 = 10.22$ and $df = 4$.

Solution: 10.22 is between 9.49 at $\alpha = 0.05$ and 11.14 at $\alpha = 0.025$. The reported P value is between $2\frac{1}{2}\%$ and 5%.

Sample Problem and Answer: Find the reported P value for a hypothesis Test of Homogeneity if $X^2 = 8.05$ and $df = 2$.

Solution: 8.05 is between 9.21 at $\alpha = 0.01$ and 7.38 at $\alpha = 0.025$. The reported P value is between 1% and $2\frac{1}{2}\%$.

CHAPTER 14 DISCUSSION QUESTIONS

These questions may be used in your study group or simply as topics for individual reflection. Whichever you do, take time to explain verbally each topic to insure your own understanding. Since these questions are intended as topics for discussion, answers to these questions are not provided. If you find that your are not comfortable with either your answers or that your group has difficulty with the topic, take time to meet with your professor to get help.

1. What is the difference between a test for goodness-of-fit and a test of independence?

2. What is the difference between a test for goodness-of-fit and a homogeneity test?

3. What is the difference between a test of independence and a homogeneity test?

4. What is the minimum expected frequency for any cell?

5. What is the relationship between independent characteristics and independent events?

6. How does one find the expected frequencies for the cells for tests of independence and homogeneity tests?

7. What is the difference between observed frequencies and expected frequencies?

8. What is a contingency table?

9. How does one find degrees of freedom for the chi-square test?

10. How does the shape of the chi-square curve for $df = 1$ or $df = 2$ differ from the shape of the distribution for other values of degrees of freedom?

CHAPTER 14 TEST

To determine if three comparable restaurants are equally popular, a survey was done during a Friday evening to find the number of people at each restaurant. The results were 28 in restaurant 1, 52 in restaurant 2, and 49 in restaurant 3.

1. State the null and alternate hypothesis for a test of goodness of fit.

2. Find the degrees of freedom and critical value for χ^2 at $\alpha = 5\%$.

3. Find E, the expected frequency, for each cell.

4. Find the value of X^2 and state your conclusion.

At one of the restaurants during a weekend, the manager surveyed the customers regarding their preference of music:

	Easy Listening	Hard Rock	Country Western	Total
Men	17	19	13	49
Women	16	12	23	51
Totals	33	31	36	100

5. State the null and alternate hypothesis for a test of independence.

6. Find E, the expected frequency, for each cell.

7. Find the X^2 value and state your conclusion at $\alpha = 10\%$.

A marketing consultant is hired by the three restaurants to determine the age of the customers. The results are:

Restaurant:	1	2	3	Total
Under 30	10	30	12	52
30 and Over	18	22	37	77
Totals	28	52	49	129

8. State the null and alternate hypothesis for the homogeneity test.

9. Find E, the expected frequency, for each cell.

10. Find the X^2 value and state your conclusion at $\alpha = 1\%$.

Chapter 14 Test Questions and Answers

To determine if three comparable restaurants are equally popular, a survey was done during a Friday evening to find the number of people at each restaurant. The results were 28 in restaurant 1, 52 in restaurant 2, and 49 in restaurant 3.

1.   State the null and alternate hypothesis for a test of goodness of fit.

Answer: $H_0: p_1 = p_2 = p_3$ $H_a: H_0$ is not true.

2.   Find the degrees of freedom and critical value for χ^2 at $\alpha = 5\%$.

Answer: Degrees of freedom $= k - 1 = 2$ $\chi^2(0.05, 2) = 5.991$.

3.   Find E, the expected frequency, for each cell.

Answer: $E = np = 129\left(\frac{1}{3}\right) = 43$.

4.   Find the value of X^2 and state your conclusion.

Answer: $X^2 = \dfrac{(28 - 43)^2}{43} + \dfrac{(52 - 43)^2}{43} + \dfrac{(49 - 43)^2}{43}$

$\qquad = \dfrac{225}{43} + \dfrac{81}{43} + \dfrac{36}{43} = \dfrac{342}{43} = 7.953$.

The value of X^2 is larger than the critical value of 5.991, so we reject the null hypothesis that the restaurants are equally popular.

At one of the restaurants during a weekend, the manager surveyed the customers regarding their preference of music:

	Easy Listening	Hard Rock	Country Western	Total
Men	17	19	13	49
Women	16	12	23	51
Totals	33	31	36	100

5. State the null and alternate hypothesis for a test of independence.

Answer: H_0: Gender and music preference are independent.
H_a: Gender and music preference are related.

6. Find E, the expected frequency, for each cell.

Answer:

	Easy Listening	Hard Rock	Country Western
Men	$\dfrac{49(33)}{100} = 16.17$	$\dfrac{49(31)}{100} = 15.19$	$\dfrac{49(36)}{100} = 17.64$
Women	$\dfrac{51(33)}{100} = 16.83$	$\dfrac{51(31)}{100} = 15.81$	$\dfrac{51(36)}{100} = 18.36$

7. Find the X^2 value and state your conclusion at $\alpha = 10\%$.

Answer: $X^2 = \dfrac{(17-16.17)^2}{16.17} + \dfrac{(19-15.19)^2}{15.19} + \dfrac{(13-17.64)^2}{17.64} +$

$\dfrac{(16-16.83)^2}{16.83} + \dfrac{(12-15.81)^2}{15.81} + \dfrac{(23-18.36)^2}{18.36}$

$X^2 = 0.04 + 0.96 + 1.22 + 0.04 + 0.92 + 1.17 = 4.350.$

The value of X^2 is smaller than the critical value of $\chi^2(0.10,(3-1)(2-1)) = 4.605$, so we fail to reject the null hypothesis that gender and music preference are independent.

A marketing consultant is hired by the three restaurants to determine the age of the customers. The results are:

Restaurant:	1	2	3	Total
Under 30	10	30	12	52
30 and Over	18	22	37	77
Totals	28	52	49	129

8. State the null and alternate hypothesis for the homogeneity test.

Answer: H_0: The three populations are homogeneous with respect to the proportion of customers under 30.
H_a: The three populations are not homogeneous with respect to this characteristic.

9. Find E, the expected frequency, for each cell.

Answer: $E_{1,1} = \dfrac{52(28)}{129} = 11.287 \qquad E_{2,1} = \dfrac{77(28)}{129} = 16.713$

$E_{1,2} = \dfrac{52(52)}{129} = 20.961 \qquad E_{2,2} = \dfrac{77(52)}{129} = 31.039$

$E_{1,3} = \dfrac{52(49)}{129} = 19.752 \qquad E_{2,3} = \dfrac{77(49)}{129} = 29.248$

10. Find the X^2 value and state your conclusion at $\alpha = 1\%$.

Answer: $X^2 = $ $\dfrac{(10-11.287)^2}{11.287} + \dfrac{(30-20.961)^2}{20.961} + \dfrac{(12-19.752)^2}{19.752} +$

$\dfrac{(18-16.713)^2}{16.713} + \dfrac{(22-31.039)^2}{31.039} + \dfrac{(37-29.248)^2}{29.248}$

$X^2 = $ $0.147 + 3.898 + 3.042 + 0.099 + 2.632 + 2.055 = 11.873$.

The degrees of freedom are $df = (3-1)(2-1) = 2$.

$$\chi^2(0.01,2) = 9.210$$

The value of X^2 is larger than the critical value, so we reject the null hypothesis and conclude that the three populations are not homogeneous with respect to the proportion of customers under the age of 30.

$\sum{}_{\circ}$

chapter fifteen

ANALYSIS OF VARIANCE (ANOVA)

Variability is the law of life.

– Sir William Osler

INTRODUCTION

This chapter is a natural extension of the hypothesis tests we have done using inferential statistics. The use of the standard score as a test statistic with single samples in chapter 9 was focused on the difference of the sample mean and population mean. However, the value of the test statistic was really dependent on the value of the standard deviation, and hence the variance. As we extended the hypothesis tests to small samples in chapter 10 and independent samples in chapter 11, the focus on means was subtly changed. When the sample or samples were small, we had to consider the effect that variance had on the value of the test statistic. In essence our hypothesis tests involved an analysis of variance. In fact, most of the hypothesis tests we have done up to this point are special cases of a more general analysis that focuses on variance. The generalization is called analysis of variance, or **ANOVA**.

The advantage of using ANOVA is that it allows us to extend the experimental design of a control group versus a treatment group to a design that considers different levels of treatment. With ANOVA, we test a null hypothesis that all of the population means are equal against an alternate hypothesis that there is at least one mean not equal to the others. It might, at first, seem reasonable to assume that ANOVA involves repeated experiments that compare all of the treatments, two at a time. However, the repeated experiment design would create a problem with the size of the type I error when we reject H_0. If we use repeated experiments, the probability of rejecting at least one null hypothesis when the population means are equal would become unacceptably large. The ANOVA design actually involves all levels of treatment as part of one experiment.

We find the sample mean and variance for each of the k levels of treatment. Using these values, we obtain two different estimates of the population variance. The first estimate of the population variance is obtained by finding the sample variance of the k sample means from the **grand mean**. This **between sample estimate of** σ^2 is referred to as the variance between the means. The second estimate of the population variance is found by using a weighted average of the sample variances. This **within sample estimate of** σ^2 is called the variance within means.

The two estimates of the population variance are then compared using the F test, with F equal to **Mean Squares Between (Treatment)** divided by the **Mean Squares Within (Error)**. This calculated value of F is then compared to a critical value of F at $n - k$ and $k - 1$ degrees of freedom to determine if we should reject the null hypothesis. When there is no treatment effect, the ratio should be close to 1. If there is a treatment effect, the observed value of F will be greater than the critical value. We then reject the null hypothesis and argue that at least one of the treatments was effective.

The F test assumes normal populations with equal variance, and independent samples. Analysis of variance is usually done using an **ANOVA table** that is generated by a computer. The analysis is sensitive to inequality of variance (**heteroscedasticity**) when the sample sizes are small and unequal and care should be used in interpreting the results.

CHAPTER 15 HOT SPOTS

If you find other HOT SPOTS, write them down and use them as a focus of your discussions in the study group. Or you can use the HOT SPOT as the topic of a help session with your professor.

Hot Spot #1 – The *t* Test Versus ANOVA

When we have two treatment levels, the *t* test used with independent samples is just a special case of ANOVA. The following example illustrates the point:

Treatment 1   Treatment 2

$\bar{x}_1 = 57.3$ $\bar{x}_2 = 54.7$

$s_1^2 = 12.4$ $s_2^2 = 12.6$

$n_1 = 25$ $n_2 = 25$

Test $H_0 : \mu_1 = \mu_2$ and $H_a : \mu_1 \neq \mu_2$ at $\alpha = 5\%$.

Using the *t* test for independent samples with variances equal and the short cut for sample sizes equal,

$$t = \frac{57.3 - 54.7}{\sqrt{\dfrac{12.4}{25} + \dfrac{12.6}{25}}} = \frac{2.6}{\sqrt{\dfrac{25}{25}}} = 2.6$$

The observed value of $t = 2.6$ is in the rejection region determined by the critical values of $t\left(\frac{0.05}{2}, 48\right) = \pm 2.0106$. Therefore, we reject H_0 and state that there is a significant difference in population means.

Using the *F* test for ANOVA we do the following analysis:

$$\bar{x} = \frac{(25)(57.3) + (25)(54.7)}{25 + 25} = 56$$

$$MS_T = \frac{(25)\left((57.3 - 56)^2 + (54.7 - 56)^2\right)}{2 - 1} = 84.5$$

$$MSE = \frac{(24)(12.4 + 12.6)}{25 + 25 - 2} = 12.5$$

Using $F = \dfrac{MS_T}{MSE}$, we get $F = \dfrac{84.5}{12.5} = 6.76$

The observed value of $F = 6.76$ is in the rejection region determined by the critical value of $F(0.05, 1, 48) = 4.0427$. Therefore, we reject the hypothesis that the population means are equal.

We now compare the results of the two hypothesis tests:

t Test	ANOVA
$\alpha = 0.05$	$\alpha = 0.05$
$t = 2.6$	$F = 6.76$
$t\left(\frac{0.05}{2}, 48\right) = \pm 2.0106$	$F(0.05, 1, 48) = 4.0427$

Table 15–1:
Comparison of
Two Hypothesis
Tests: t Test
Versus ANOVA

The point of the above comparison is that if we square both the observed t value and the critical value of t, we get the corresponding values of F we have in the ANOVA column. The F test for two treatment levels is simply the square of the t test for independent samples. The critical values are related using the formula:

$$t^2\left(\frac{\alpha}{2}, n - 2\right) = F(\alpha, 1, n - 2)$$

Thus, the t test we used with two independent samples with population variances equal, is a special case of ANOVA with $k = 2$ treatment levels. There are three advantages to using ANOVA:

1. We have an experimental design that will work with more than two treatment levels.
2. The F test is not as sensitive as the t test to errors that occur in rounding off values for the mean and standard deviation.
3. We are able to focus on the F test rather than having to know and remember both the t and F test statistics.

The above comparison between the t test and ANOVA was done using two groups of equal sample size. This was done to simplify the comparison. However, ANOVA can be used with treatment levels involving groups of unequal sample size.

Analysis Of Variance (ANOVA)

In order to use ANOVA, we need independent random samples from normal populations with equal variances. If these assumptions are violated and the sample sizes are both unequal and small, we can use an analysis, such as the Kruskal-Wallis test, from nonparametric statistics.

Johnson And Bhattacharyya Section 15.2

Hot Spot #2 – The Importance Of The ANOVA Table

There is an obvious advantage to using a computer to do the number crunching that is involved with large samples. As the number of treatment levels increase, it is imperative that the computer be used to avoid errors in the analysis of the data. Most of the computer outputs for analysis of variance are done with the ANOVA table. As such, it is worthwhile to learn how to read an ANOVA table. However, there is an even more important reason for becoming familiar with ANOVA tables.

Many of the statistical software packages used on the computer treat bivariate data analysis (Regression and Correlation) as a special case of multivariate statistical analysis. The output of these programs include the value of the t test statistic for the null hypotheses that the coefficients of the independent variables are each zero, the level of significance for each t test, and the ANOVA table for the null hypothesis that the population correlation coefficient or multiple correlation coefficient is zero.

The ANOVA table becomes a common structure for both Analysis of Variance and Multivariate Data Analysis that includes bivariate regression and correlation. Because of this it is important to

1. become familiar with the ANOVA table,
2. become comfortable with the terminology,
3. know how to interpret the results.

CHAPTER 15 DISCUSSION QUESTIONS

These questions may be used in your study group or simply as topics for individual reflection. Whichever you do, take time to explain verbally each topic to insure your own understanding. Since these questions are intended as topics for discussion, answers to these questions are not provided. If you find that you are not comfortable with either your answers or that your group has difficulty with the topic, take time to meet with your professor to get help.

1. What do the letters ANOVA represent?

2. What is the null hypothesis using ANOVA?

3. Why do we use ANOVA for k treatments rather than doing all possible comparisons two at a time?

4. What assumptions are made when using ANOVA?

5. How are dot diagrams of residual plots used to examine the assumptions involved using the ANOVA model?

6. Explain the decomposition of the degrees of freedom associated with the sum of squares in ANOVA.

7. Give the formula for the decomposition of the total sum of squares into sum of squares treatment and sum of squares error?

8. How do the alternate formulas for ANOVA avoid the round off errors associated with calculations of the mean and standard deviation?

9. What information is included in an ANOVA table?

10. What are multiple-t confidence intervals?

CHAPTER 15 TEST

Given the following information, use ANOVA:

Treatment 1	Treatment 2	Treatment 3
$\bar{x}_1 = 56.1$	$\bar{x}_2 = 54.9$	$\bar{x}_3 = 52.4$
$s_1^2 = 10.8$	$s_2^2 = 10.2$	$s_3^2 = 10.5$
$n_1 = 20$	$n_2 = 20$	$n_3 = 20$

1.   State the null and alternate hypothesis.

2.   Find the grand mean $\bar{x}$.

3.   Find MS_T.

4.   Find MSE.

5.   What is the observed value of the F statistic?

6.   What is the critical value of F? State your conclusion.

Given the following ANOVA table:

Source	Sum of Squares	df	MS	F
Between Samples	112	2		
Within Samples	238	27		

7.   Find MS_T.

8.   Find MSE.

9.   What is the observed value of the F statistic?

10.   What is the critical F value at $\alpha = 5\%$? State your conclusion.

CHAPTER 15 TEST Questions and Answers

Given the following information, use ANOVA:

Treatment 1	Treatment 2	Treatment 3
$\bar{x}_1 = 56.1$	$\bar{x}_2 = 54.9$	$\bar{x}_3 = 52.4$
$s_1^2 = 10.8$	$s_2^2 = 10.2$	$s_3^2 = 10.5$
$n_1 = 20$	$n_2 = 20$	$n_3 = 20$

1.   State the null and alternate hypothesis.

Answer: $H_0: \mu_1 = \mu_2 = \mu_3$ and $H_a:$ not all means equal.

2.   Find the grand mean $\bar{x}$.

Answer: $\bar{x} = \dfrac{(20)(56.1) + (20)(54.9) + (20)(52.4)}{20 + 20 + 20} = 54.47$.

3.   Find MS_T.

Answer: $MS_T = \dfrac{(20)\left((56.1 - 54.47)^2 + (54.9 - 54.47)^2 + (52.4 - 54.47)^2\right)}{3 - 1}$

$MS_T = \dfrac{(20)(7.12)}{3 - 1} = 71.2$.

4.   Find MSE.

Answer: $MSE = \dfrac{(19)(10.8 + 10.2 + 10.5)}{20 + 20 + 20 - 3} = 10.5$

5.   What is the observed value of F?

Answer: $F = \dfrac{MS_T}{MSE} = \dfrac{71.2}{10.5} = 6.78$.

Analysis Of Variance (ANOVA)

6.   What is the critical value of F? State your conclusion.

Answer: $F(0.05, 3 - 1, 60 - 3) = F(0.05, 2, 57) = 3.159$.
         Conclusion: Reject H_0.

Given the following ANOVA table:

Source	Sum of Squares	df	MS	F
Treatment	112	2		
Error	238	27		

7.   Find MS_T.

Answer: $MS_T = \dfrac{SS_T}{k-1} = \dfrac{112}{2} = 56$

8.   Find MSE

Answer: $MSE = \dfrac{SSE}{n-k} = \dfrac{238}{27} = 8.81$

9.   What is the value of the F statistic?

Answer: $F = \dfrac{MS_T}{MSE} = \dfrac{56}{8.81} = 6.36$

10.   What is the critical F value at $\alpha = 5\%$? State your conclusion.

Answer: $F(0.05, 3 - 1, 30 - 3) = F(0.05, 2, 27) = 3.35$.
         Conclusion: Reject H_0.

$\sum °$

chapter sixteen

NONPARAMETRIC INFERENCE

Mathematicians are like Frenchmen;
Whatever you say to them,
They translate it into their own language
And forthwith it is something entirely different.

– Goethe

INTRODUCTION

Throughout our discussions of hypothesis testing we have made assumptions about the distribution of the variable involved. Generally speaking the assumptions have been that we had a normal distribution, (bell shaped curve). Most of the time, however, the normal curve only approximates our distribution. But we then use the value of a test statistic such as z, t, F, and χ^2 to obtain an exact value for the probability that the test statistic is in a particular interval.

There is a natural question that needs to be addressed, namely what happens if our assumptions are not valid. In other words, what do we do if we do not have a bell shaped distribution. The solution to the problem is to use what are called **nonparametric statistics** where we make minimal assumptions about the distribution.

There are both advantages and disadvantages to using nonparametric statistics. The first advantage is that the mathematics involved is usually very simple and easy to understand and only requires that the variable is measured using an interval scale or ratio scale. The most important advantage is that the nonparametric test may be more powerful than the parametric test when the assumption of normality is not maintained. The **power of a test** is $1 - \beta$, where β is the probability of failing to reject a false null hypothesis. Thus, the power of a test is the probability of rejecting a false null hypothesis.

There are, however, some disadvantages to using nonparametric tests. Often times the parametric test may be more powerful; even when we cannot maintain the assumption of a bell shaped distribution. Furthermore, even though the nonparametric method is an exact model of our distribution, our answers for the probabilities involved are only approximations because of the information we must ignore when we use nonparametric methods.

Given the advantages and disadvantages, it is best to simplify the issue as follows. If it is very clear that our distribution is not a bell-shaped, **symmetric distribution,** and our sample size is small, then we should use a nonparametric or **distribution free method**. If our sample size is large and the distribution is still roughly bell- shaped, we should stay with parametric methods.

There is the additional advantage to parametric statistics in that they are usually better known than the nonparametric methods. As a result the analysis we go through may be more readily accepted when we use parametric methods.

The nonparametric tests covered in this chapter are the **Wilcoxon Rank Sum Test**, the **Sign Test**, the **Wilcoxon Signed-Rank Test**, and the **Spearman's Rank Correlation Test**.

CHAPTER 16 DISCUSSION QUESTIONS

These questions may be used in your study group or simply as topics for individual reflection. Whichever you do, take time to explain verbally each topic to insure your own understanding. Since these questions are intended as topics for discussion, answers to these questions are not provided. If you find that your are not comfortable with either your answers or that your group has difficulty with the topic, take time to meet with your professor to get help.

1. What is a nonparametric test?

2. What is the difference between a parametric and nonparametric test?

3. What is a distribution free test?

4. When are parametric tests better than nonparametric tests?

5. What is the power of a test?

6. When should a nonparametric test be considered?

7. If two statistical tests are used at the same level of α, how does one decide on the better test?

8. What is the advantage of using Spearman's Rank Correlation, r_{sp}, over Pearson's correlation coefficient, r?

9. What are the nonparametric tests that can be used for comparing two treatments?

10. What are the nonparametric tests that can be used for independent samples?